Sensor Application in Managing in-season Crop Variability

Sensor Application in Managing in-season Crop Variability

Solange Uwera

Delve
PUBLISHING

www.delvepublishing.com

Sensor Application in Managing in-season Crop Variability

Solange Uwera

Delve Publishing
224 Shoreacres Road
Burlington, ON L7L 2H2
Canada
www.delvepublishing.com

Email: orders@arclereducation.com

ISBN: 978-1-77469-038-3 (Hardcover)

ABOUT THE AUTHOR

Solange Uwera is a GIS and Remote Sensing Expert. She received her Master's degrees in Geographical Information Science from the University of Manchester and in Human Settlement and Urbanism from the University of Leuven and a bachelor's degree in Biodiversity Conservation from the National University of Rwanda. Solange works on various development and humanitarian projects at United Nations Satellite Centre. She worked in Geneva, Brussels, Rwanda and Madagascar where she applied geospatial technology for humanitarian, development and environmental applications.

TABLE OF CONTENTS

LIST OF FIGURES

PREFACE

Climate change has changed the course of agricultural practice these days. Delayed attention was grasped on this essential concept per se. It has already started exhibiting several problems in the natural ecosystems thereby altering the course of our understanding of the climate and its potential. Most human societies are on the verge of vulnerability because of anthropogenic interventions on the biotic and abiotic factors. It is to be noted that climate is not just an average of the weather events but it is the mirror that helps us understand the future possible scenarios of the climate aberrations. We are witnessing uneven floods here and there and this can be attributed to the extreme climatic paradigm shift. It is obvious that almost all the climatologists and meteorologists along with the common people know the effects of El Nino and Southern Oscillations. This was studied extensively as it regulates climate and weather patterns all over the world. The climatic variability can be due to various factors and they drive the dynamics of natural ecosystems and agricultural production. We can use geographic information systems or GIS to generate maps that can project the climate scenarios.

In this book, the sensor applications are explained with respect to the agricultural sectors. The discussion was made about the sensors and platforms. Remote sensing and GIS are considered as important tools in mapping agricultural fields and their associated factors. The mapping of crops with several space-borne platforms and airborne platforms is explained in this book. The role of climate variability on the crops is discussed with contemporary examples and applications. Models play a vital role in predicting the risks associated with agriculture and allied fields. In order to describe the role of models played in climate forecast, a chapter is dedicated to explaining the climate system models. The essential role of agrometeorology on crops is explained in this book. Cool-season legumes constitute the majority of the agricultural products and climate management pertaining to the growth of these cold legumes is discussed in detail. The role of geosensors is an important aspect to discuss and this book can help us to know the potential applications of these geosensors. We can use these sensors in the crop variability studies. A detailed account of hyperspectral remote sensing, UAVs and seasonal variability is given in this book.

This book helps as a guide to all the students who would like to help the farming communities with ease. This book can be used as a handbook for engineers, scientists and research scholars along with undergraduates and postgraduates in science and technology.

Introduction

Climate change can be linked with the seasonal variations that might have occurred in the past decades and were recorded with reliable instruments. The climate extremes can bring about several changes in our lifestyle as the basics of our existence is becoming unconducive with the shortage of food products. There is always a demand for research works that are linked with modeling of the agricultural landscape along with the factors that enclose the food production land parcels. The sensors can be classified as advanced tools that can help us capture the essential information needed for us to perform agriculture-related climate forecasts. These tools can give us an edge to improve crop production capabilities thereby limiting the factors that lower the crop yields. The trends obtained or followed by the agricultural yields can be mapped with the aid of remote sensing and GIS technology. We can know the potential threats that might induce damage to the crops. There are many reasons to study climate variabilities as most of the extremes can affect the agricultural sector by and large. We need to consider both temperature and oceanic-atmospheric circulations so as to arrive at a meaningful conclusion. Most of the agricultural fields are prone to climate-induced vulnerabilities and rainfall over a region can be classified as an essential entity in this case. It is obvious that we need to predict the seasonal climate with precision and this should be backed with many tools.

It is essential to produce climate system models that can bring about an accurate representation of the climate forecast. The concept of global warming has become a local phrase for every region and its effects are obvious on the farming communities. The models generated can be used to project the current situations and can help us predict or forecast the potential perils in the future. Most of the crops that are grown are considered sensitive to the meteorological conditions and the factors that might control these conditions should be studied in detail so as to generate accuracy and precision. This aids us in producing higher yield rates. The cold legumes can help us improve human health and it is essential to understand the growth rates of these cold legumes in relation to the climate schematics. Most of these legumes have adaptation capacity but some are still in the process of yielding adaptation appendages. With the surfacing of advanced technologies in the sensors, the latest amalgamation of geoinformation with sensors was generated and they were named as geosensors. These sensors can feed us with real-time data and the end-users will enjoy the privilege of datasets with extreme accuracy and resolutions. In order to reduce the costs incurred in the application of technological inputs in agriculture, unmanned aerial vehicles were constructed and they come up with immense possibilities as far as agriculture-related sectors are concerned.

Chapter 1

Introduction to Sensor Applications

CONTENTS

1.1 INTRODUCTION

Climate change has changed the seasonal variations for decades and most of the crops are being affected. It is important to enhance the techniques needed in the weather forecasting methodologies. Several statistical methods were used to forecast seasonal variabilities. Most of the climate extremes can lower the crop production rates and due to this, the food exports dwindled (Headey et al., 2011). Most of the countries are based on food exports and imports. There is a demand for forecasting the climate disparities so that the agriculture sector can be protected from weather-based perils. The forecast models have become popular in almost every country due to their accuracy and advantages. The monitoring of the crops can be managed easily with the aid of the sensors. The use of drones and satellite sensors has become popular. The negative impacts of climate disparities can be handled effectively with climate forecasts. The crop yield forecasting services can supply data to certain regions like China, the United States, Africa and Europe. It is always important to use global yield models than the local models (Lizumi et al., 2001). In most cases, we can use multi-model ensemble techniques to enhance the dependability of climate and weather forecasts over spatial-temporal domains (Knutti et al., 2011). As per an estimate, the need for agricultural products can enhance by the year 2050 (Godfray et al., 2014). There are many problems whilst enhancing crop yields. The arable land per capita is being decreased every day and this can be attributed to climate change (FAO, 2000). It is obvious that the climate information like seasonal forecasting of ENSO or El Nino Southern Oscillation has helped farmers to know the risks (Bussay et al., 2015). The ENSO phenomena can affect agriculture all over the world and this can be due to the onset of hydro-meteorological deviations (Casanueva et al., 2006). The ENSO phenomena can also affect climate extremes (Halide et al., 2017). This phenomenon can affect the crop yields directly and indirectly and so there will be buoyancy in the prices of the agricultural produce (Ray et al., 1987). It also alters the climate conditions in European countries (Bronnimann et al., 2002). Remote sensing of crops from the aircraft and satellites can give us information about several parameters. The data obtained via these instruments can be useful to predict the risks. We can use the crop data along with the weather forecasting models to arrive at a conclusion regarding the factors that influence crop growth patterns.

In this decade, there were several advancements in precision agriculture that demanded the use of drones and other data collection devices. These devices can also monitor crop growth and pathology. The farming technique can yield many challenges as far as productivity is concerned. The population growth and urbanization along with the deteriorating environment are putting pressure on the natural resources that are already scarce. This pressure is immense on the agriculture and allied sectors. Climate change has become the biggest problem when it comes to food production.

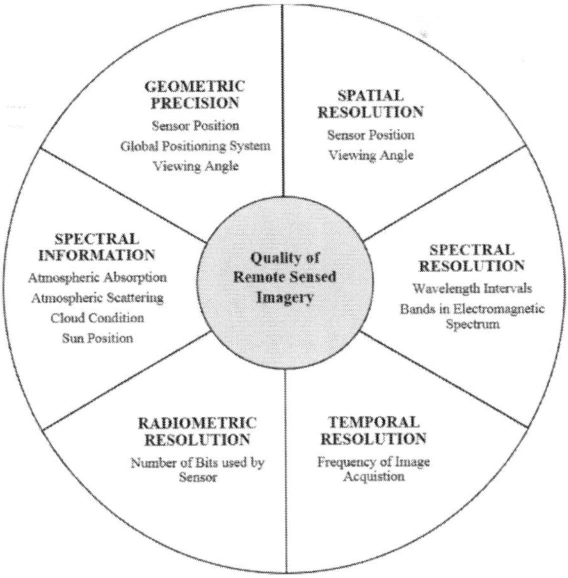

Figure 1.1: Sensor applications in Agriculture.

Source: https://ohioline.osu.edu/sites/ohioline/files/imce/Food_Agricultural_and_Biological_Engineering/FABE-554.1__Fig1-Factors-influencing-quality-images.jpg

The crop yields can be based on the growth variabilities and these changes can be based on several factors. The distribution of water in the fields can be considered as an essential factor in determining the crop yield and variability. If the water supply is abundant, then we can expect better yields. We can model the water distribution in the crop fields using many software. We can use the topographic wetness index or TWI for modeling the water distribution (Sorensen et al., 2006). We can consider the canopy reflectance to understand and evaluate the crop yields in a particular season. There are several vegetation indices that can help us to understand the crop variability and their dependence on several external factors. There are different types of sensors that can be used for this purpose. These sensors mostly measure the reflected solar radiation. Landsat 7 was used for this purpose since its introduction into space. The Landsat 7 data comprise of 7 bands. The resolution offered by these images is less and this can be considered as a major problem with these products. Latest satellite platforms include WorldView-2, QuickBird and others that provide high-resolution data (Mulla, 2014). Most of the literature suggests that the remote sensors can be important in evaluating crop yields. Climate change can be classified as one of the reasons behind threatened food security all over the world. This is common in the sub-Saharan regions (Eva, 2013). These countries food sources are mainly local agricultural products. It is affirmative to say that global warming has been increased and it can be attributed to precipitation, emission of carbon dioxide and other factors (Fraser, 2019). In Eastern parts of Africa, we can observe intense rainfall that can devastate the crops (Michael, 2006). It is obvious through several observations that the erratic distribution and insufficient rainfall can affect the crops drastically. The income generation capacity of the small and marginal farmers will be reduced and this situation can be attributed to the climate change patterns.

The reflectance sensors (Figure1.1) can be used to evaluate the nitrogen requirements of the crops. These sensors can offer us several advantages economically and technically. These sensors can cover most of the areas and can bring us effective results. We can also use passive sensors for this purpose that can sense deviations from normal solar radiation. Active sensors need energy and so we can rely on them to extract the data of our needs. These sensors can collect data in any weather condition. The agricultural sector can be considered an extremely sensitive domain as far as the human population is concerned. Due to climate change, the growing season will be prolonged and this can be classified as an anomaly. Agriculture can be important for both poor and middle-income communities. Most of these communities fall apart if the present scenario continues.

Precision agriculture (Figure 1.2) is one such concept these days that can enhance crop yields drastically. It manages the effect of agriculture on the environment. This can be achieved by developing site-specific crop management. This concept is mentioned in several research papers as SSCM. This can be studied as a portion of precision agriculture. The main focus of SSCM is to manage crop yield variability with a special focus on the site potential. We can divide the cropping fields into zones. The zonation can be done based on the limiting factors of the yield. The spatial variability in the crop yield can be dependent on several factors that can regulate crop growth. Some of them are climatic factors, topographical factors and soils. The soil-based factors are the availability of soil water and nutrients. The topographic factors include elevation and other factors. Temperature and rainfall can be classified under climatic factors. The climatic factors can suffer from temporal variability. This is the reason why the crop yields differ with seasons. It is always important to understand farm management clearly so that we can decide the steps that are to be taken to abate the unequal crop yields and anticipated economic losses. The extent of variability can be considered as an important issue in integrated crop management. The SSCM concept may demand an exhaustive understanding of spatial and temporal dimensions in relation to crop growth.

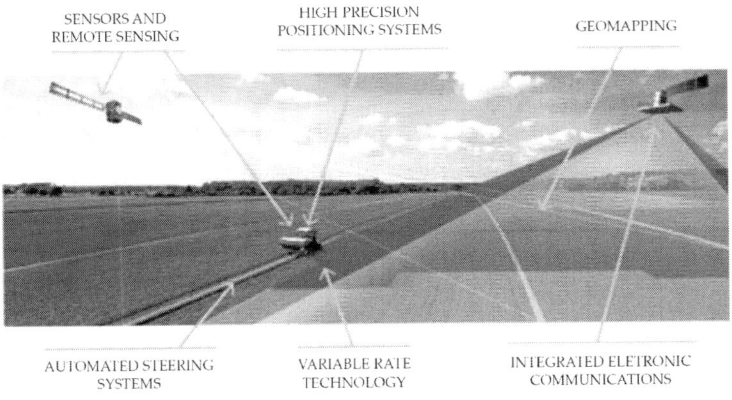

Figure 1.2: Precision Agriculture.

Source: https://www.google.com/url?sa=i&url=https%3A%2F%2Fwww.ias-express.net%2Fprecision-farming-in-india-features-merits-demerits-and-chal lenges%2F&psig=AOvVaw0bJYhjGGqoFfZLwKxsJjW0&ust=160630952892 7000&source=images&cd=vfe&ved=0CAIQjRxqFwoTCPiKjbygm-0CFQA-AAAAdAAAAABAJ

In South Australia, most of the crop yields are regulated by the availability of soil water. The available plant water holding capacity or PAWC can be important in the zonation process (Oliver et al., 2009). PAWC affects the crop yield and it is primarily based on the precipitation patterns. Whilst these studies, the spatial and temporal variograms were used to understand the temporal lag and spatial lags. This has helped us to develop maps pertaining to spatial and temporal trends. These trends can be evaluated based on multi-year data available with the national space agencies and agricultural departments.

1.2 SENSORS

Satellite remote sensing system is earth observation technology which has generated much data pertaining to agriculture. This data helped many to frame models that can enhance crop yields. Remote sensing data should be processed prior to its utilization. We can use raw images supplied from the sensors onboard aircraft and sensors. The data obtained can suffer from deformations. These deformations are due to terrain profiles and atmospheric conditions existing in the area of interest. Data obtained from diverse sources needs to be subjected to geometric and radiometric corrections. The radiometric correction can be associated with the sensitivity of the sensors, absorption, atmospheric scattering, sun angle, topography and other related parameters (Figure 1.3).

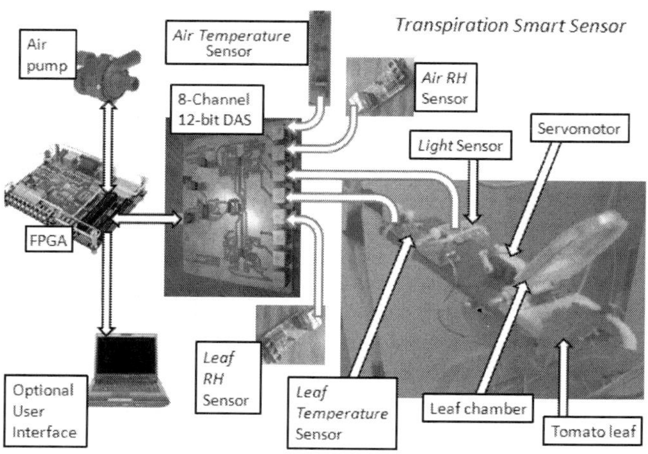

Figure 1.3: Sensors for crops.

Source: https://www.google.com/url?sa=i&url=https%3A%2F%2Fwww.mdpi.com%2Fjournal%2Fsensors%2Fspecial_issues%2Fsensor_agriculture%3Fview%3Dabstract%26listby%3Ddate&psig=AOvVaw3NkA8adAdDldTCZBYjwNgU&ust=1606309712521000&source=images&cd=vfe&ved=0CAIQjRxqFwoTCNDA8fygm-0CFQAAAAAdAAAAABAD

The atmospheric correction can be a laborious task for agro-technicians and agronomists. This is due to the fact that data pertaining to image acquisition and atmospheric conditions are to be known. The information and data can differ with location and time. Some of the popular space programs that can be linked with agriculture are MODIS or moderate resolution imaging spectroradiometer and Landsat. The satellite remote sensing data can be classified into levels. They are 0, 1A, 1B, 2,3 and 4. The 0 level data is reconstructed data. The data at this level will be unprocessed and payload data can be achieved at full resolution. Some of the artifacts like duplicate data, communication headers and synchronization frames will be eliminated through processing. Level 1A data can be classified as reconstructed data. The data at this level will be unprocessed and this will be done at high resolution. At this level, we can time-reference the images and appropriate annotation can be done. Some of the ancillary information like geometric and radiometric calibration will be added to the images. Some of the georeferencing properties will be added to the data obtained from the sensors. From level 1A data, we can get level 0 data. Level 1B data is obtained from level 1A data and it can be processed with the aid of units pertaining to sensors like brightness temperature, backscattering and others. We cannot yield level 0 data from level 1B data and it is important to know that all the sensor systems do not have level 1B data. We can get the agro-geophysical data from level 2 data. Data pertaining to canopy temperature, soil moisture, ice concentration and height of ocean waves can be derived from level 2 data. Level 3 data can be used to map the variables at equal spatial grids. We can get consistency and completeness of data at this level. We can interpolate with this data. Level 4 data can be used in modeling of the data and it is directly linked with the measurements. EOS-TERRA/AQUA satellites carry MODIS (Figure 1.4) and all levels of data are available with this payload. We can get the data in HDF-EOS formats. This satellite was produced by the United States. Landsat 7 has ETM+ or Enhanced Thematic Mapper Plus as payload. This has data classified as levels 0,1,2,3 and 4. The data obtained can be in HDF format. Landsat 8 has OLI and TIRS as payloads. The data obtained from this sensor will be radiometrically and geometrically rectified. We can get data in HDF format. Quickbird is another important satellite that can offer high resolution. The payload has panchromatic and multispectral sensors. We can get geometrically rectified and radiometrically rectified data with these sensors. We can get the data in the form of GeoTIFF format. GeoEye-1 is another important satellite that can provide us both multispectral and panchromatic images. We can get data at levels 0,1 and 2. The products can be subjected to geoprocessing with the aid of softwares. The images can be in GeoTIFF formats.

The remote sensing technology started from the event of aerial photography. The remote sensing techniques with the help of satellites came into force in the 1960s. There are several advances in this field since then. Some of the advances are UAV, airborne remote sensing and Lidar techniques. The remote sensing products that are obtained from UAV can monitor agricultural lands on a real-time basis. We can combine images from multiple phases, scales and sources in the process of image fusion. This image fusion can help us in image analysis. We can fuse images with higher resolution with that of low-resolution images. This process can be called pan-sharpening. This data fusion technique can be performed at the decision level and feature level mostly (Andrea et al., 2009). The main advantage of the UAV remote sensing systems is that they can be operated at low altitudes thus enabling us to try precision agriculture. Most of the UAV systems are least susceptible to weather conditions. We can use the mosaicking technique if there are multiple images. Stereo vision can be used in this case to produce three-dimensional models. Through UAV's we can get the biomass volumes and height of the plant. We can use the images obtained from the remote sensing instruments to analyze data qualitatively and quantitatively.

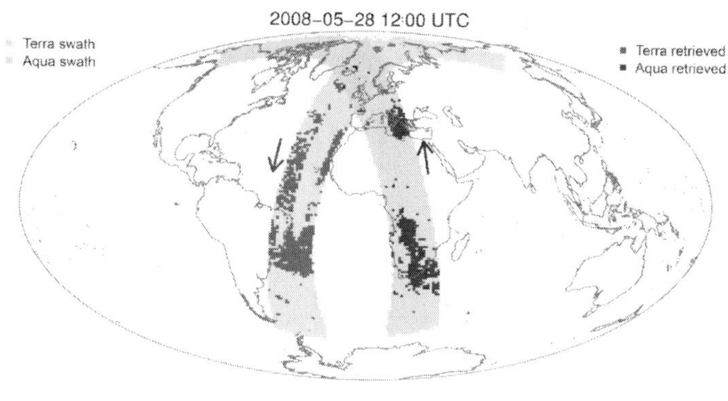

Figure 1.4: MODIS platform.

Source: https://amt.copernicus.org/articles/11/4073/2018/amt-11-4073-2018-f06-web.png

Most of the data analysis can be done with the aid of classification techniques. The classification techniques are important in precision agriculture as there will be plenty of applications that can be addressed. Most of the quantitative analysis with the data obtained can be based on modeling procedures. The analysis of remote sensing data is based on radiative transfer theory. Machine learning can be used in the classification of the images. We need to visualize the data and products that are obtained from the remote sensing systems. Visualization can be based on modularization, Web-enabling, mobility and data-based techniques. GIS and remote sensing must be integrated to get appropriate results. The spatial and temporal analysis can be considered as important as far as the remote sensing imagery is concerned. The remote sensing of agriculture can be done via airplanes and satellites that are located at different altitudes. Out of all the remote sensing devices, the low altitude remote sensing or LARS can be considered as important. The sensors that are installed on the tractors can be used for sensing the changes in the crop fields. Optimal irrigation can be considered as an important parameter that can determine crop health. The sensors can be used to estimate the moisture that is present in the soils. These sensors can be classified as essential elements in the nutrient management of the crops. Knowing the water status of the cropland can help us design the drainage requirements of the fields. This will help us to reduce the nutrient losses whilst leaching. An optimal nutrient is possible only if advanced techniques like remote sensing are to be used. With the inclusion of remote sensors in detecting several nutrient parameters, we can enhance crop productivity. The salinity concentrations in the soil can be mapped. Information and communication technology with the aid of electronics can help the growers to select the crop types and soil parameters that are to be maintained.

The moisture sensors can be used in enabling effective management of irrigation in most of the crop fields. Many of these methods can be classified as stand-alone tools. We can use these sensors along with the estimating methods proposed by the Food and Agriculture Organization or FAO. The moisture detection sensors can be used in knowing the soil water levels at various stages. We can estimate the soil water content in the limits called drier value and wetter values. These values can help us to plan an irrigation event whilst not jeopardizing the actual aim of irrigation management. These two limits can give us the volume of irrigation that is to be considered. We can use automatic sensors to know the soil water status. These sensors are capable of measuring the soil matric potential and volumetric soil water content. We can use these sensors as wetting detectors or wetting front detectors.

1.3 PRECISION AGRICULTURE

Precision agriculture can be classified as a relatively new concept that is being practiced these days. The precision farming systems can be a mix of information, communication technology with management techniques. In this system, we will use automated machinery and hence most of the labor costs will come down. With the usage of the sensing techniques, we can detect the pest population that can damage the crops and reduce the yields drastically. In the precision farming systems, their agricultural parameters will be checked properly. We can identify location-based mitigation processes that are to be practiced for better yields. We can send data that is provided by these sensors to the control station. At the control station, we can perform modeling and so the effective measures that are to be used can be decided then and there. Most of the agricultural sensors, foggers, irrigation systems, actuators are already present in the global markets.

Wireless Sensor Networks (Figure 1.5) or WSN can be defined as autonomous sensors that can be used to know the physical conditions that are existing in the crop fields. Temperature, rainfall, pressure and other parameters can be considered as physical conditions that determine the crop yields. The WSN can be classified as a gateway that can provide data to the end-users in less time possible. We can use advanced modem networks to establish a two-way network and controlling infrastructure. These infrastructures are prevalent in industries that deal with agricultural engineering and sensors. Sensors can be any device that can identify and respond to the stimulus from the external environment. Sensors can respond and record several physical parameters or conditions. The outputs generated from these sensors can be translated into user-defined language and will be supplied to the control station for further processing.

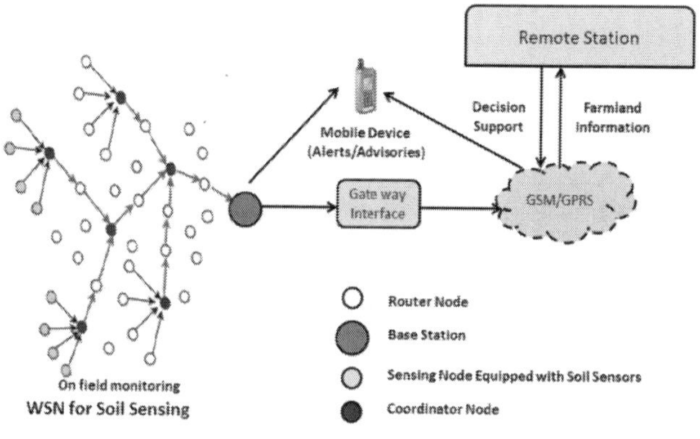

Figure 1.5: Wireless Sensor Networks.

Source: https://media.springernature.com/lw685/springer-static/image/art%3A10.1007%2Fs40012-020-00289-1/MediaObjects/40012_2020_289_Fig1_HTML.png?as=webp

The sensors that are to be used in the crop fields needs to be installed on drones, unmanned aerial vehicles (UAV) or vehicles, satellites and aircraft.

We can place the sensors in the water and soils. These days, researchers are installing sensors tagged to plants and animals. The sensors that are being used in agriculture can collect data from the irrigation systems, water pumps, weather stations, robotic dairies, weighbridges and farm machinery. We can use biosensors that are made with the aid of antibodies and enzymes along with chemicals that can react with the substances. The biosensors can detect the organophosphate residues that are present in the farm products. The bacterial population in the food produce can be detected by the biosensors. The information collected from these sensors can be called as 'big data'. This data can be analyzed and used in the improvement of crop yields. The data after processing can be used for several purposes and can be called as 'meta data'. Most of the sensors can provide us real-time information with greater accuracy and precision as and when needed. These data products can help us in making decisions that can lower the input expenditures. The sensors can be used in the primary industries. They can be used in mining and other industries that demand real-time data. CSIRO has recently manufactured a sensor technology that can be used as wearable. These sensors can analyze the health patterns in humans and are capable of sending data to the recipient. We can use sensors to analyze the conditions that exist in the greenhouse. They can be used in the horticulture field to enhance crop yields. The sensors can be used to measure the light levels, relative humidity and temperature that exist in the crop fields. The sensors can help us to maintain optimal crop conditions. The irrigation management and regulation of temperature can be considered as crucial parameters or conditions that can be maintained by the sensors. The sensors can be used to analyze every plant. The data can be transmitted continuously or in some cases in batches. The main parameters that can be recorded by the sensors are leaf temperature, fruit and flower growth rates, the diameter of the stem and other important characteristics of the crops. This data can help us to know the stress conditions of the plant.

With this data, we can proceed with the management measures. Most of the agricultural sensors can be used to analyze the growth conditions of the crop. In the case of vineyards, the growers use the weather sensors to plan their irrigation, harvest and other crop-related practices. We can even get the data about the frosts that occur overnight and the incidence of diseases. We can easily mitigate the pathogens with the help of sensors. There are many universities that have developed robotics to help precision farming. We can apply herbicides automatically in a time-phased manner. Most of the weeds will get eliminated from the crop fields. The sensors can be attached to the internal membranes of animals and microchips can have immense uses in this case. The sensors can evaluate the health of the crops and help us to decide the number of fertilizers that are to be used.

1.4 SMART AGRICULTURE

There are several efforts made in the agriculture field in order to enhance the yields according to the demands of the population. The increase in the rate of human population never gave us a chance to equate supply and demand. The increase in population is mostly seen in developing nations. The rate of urbanization is being increased and the chances of its mitigation are slim. Once the income levels of these nations become improved, people will shift to a healthier diet and this can be classified as an important problem in the coming years. It is quintessential to consider doubling the crop production yields so that the growing population might be comfortable. IoT is an important technology that can be applied to the agriculture sector. With the use of sensing techniques and IoT, the landscape of traditional farming gets changed and it is visible in most of the areas. The concept of smart agriculture is based on three important pillars called applications, services and sensors. The applications part includes monitoring of water and nutrition, monitoring of pests and diseases, monitoring of soils, crop health monitoring, environment and machinery. The services portion includes irrigation, pesticides, fungicides, herbicides, fertilization, preparation of soil, condition and storage of the yields. The sensors portion includes leaf sensors, stem sensors, temperature sensors, humidity and fruit size sensors, etc.

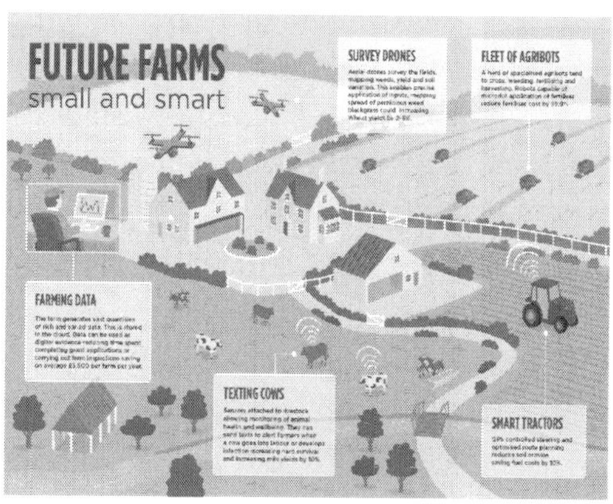

Figure 1.6: Smart agriculture.

Source: https://medium.com/@hypelabs.io/smart-agriculture-with-hype-sdk-b39769883982

The fertilization process mediated by smart agriculture (Figure 1.6) can help us in the estimation of crop requirements precisely. With the advent of IoT-based agriculture, analysis of nutrient needs for any crop spatially has become easier and less labor-intensive. We can use the Normalized Difference Vegetation Index or NDVI based on the satellite images in order to analyze the crop nutrient requirements. NDVI is based on the reflected visible and near IR radiation from the crops. This can be used to estimate the density, vigor, vegetation and health of the crops. These parameters can help us to assess the nutrient needs of the crop plants. By using this data, we can enhance the efficiency of the fertilizers and hence damage to the environment will be reduced. Variable Rate Technology or VRT, geo-mapping and GPS has become an ingredient of most sensor technologies that focus on agriculture. We can effectively go ahead with chemigation, fertigation and others by using IoT in agriculture. The inputs that can be classified as important in smart agriculture are crop data, environmental data, soil data, crop standards, weather forecasts, best practices, government policies and market demands. In the 'processes' element of smart agriculture, we can consider the preparation of land, sowing, irrigation, fertilization, pesticide and harvesting. The outputs of smart agriculture are quantity, quality, control, economy, future predictions and update requirements. In the case of yield monitoring, we can analyze several aspects that can determine the grain quantity, moisture content, grain mass and agricultural yield.

Crop forecasting is another important aspect that can estimate the yield and so the rate of production can be known beforehand. This will help the farmers in decision-making. In this process, the farm area network or FAN can be used. The direct measurement of soil chemistry can be considered as an important aspect of the sensors. We can measure temperature, pH, moisture and other essential parameters using sensors. It is always important to know the results generates whilst soil testing so as to get good yields. Ion-sensitive fields effect transistors and Ion-selective electrodes can be classified as sensors that can be used to analyze the ion uptake in plants. The uptake of nutrients can determine the growth and development of the plants. The ion-selective sensors can detect more ions than other sensors. Most of the routine agricultural operations can be based on eliminating pests and weeds along with enhanced plant growth. The amount of fertilizers and pesticides that are to be used can be standardized through the application of these sensors.

The sensors can be mostly used in detecting the changes in humidity and temperature, analysis of crop yields, crop requirements estimation, water and soil study, disease identification and fertilizer quantification. Sensors that are used in the Greenhouses can help us understand the accurate ratio of the fertilizers to be used. Most of the parameters that control the Greenhouse plants can be measured by the sensors. The acoustic sensors can be used to know the texture of the soils. We can measure the noise level fluctuations that occur in the soil particles. We can use airflow sensors to estimate the permeability of air in the soils. We can also identify the structure of the soils using these sensors. The electrochemical sensors can be used in most of the agricultural fields and they are capable of measuring the micro and macronutrients in the soils. With this sensor, we can measure pH, salinity and other essential parameters. We can map the EC or electrical conductivity of the soils using these sensors. The organic matter and nitrates that are present in the soils can be measured using these sensors. The mechanical sensors can be get penetrated into the soil and record the strain levels and strength of the soils. The canopies can be measured by the microwave and optical sensors. We can use LIDAR sensors to know the leaf area index and other parameters. GPS can be infused with this technology to get a better result. Agrosensors are being used across mobile and fixed platforms. The environmental data can be measured, processed and supplied to the ground station for analysis. We can use telemetry sensors to know the conditions that persist in the inaccessible regions.

Crop failure can be considered as an important factor behind the reduction in the annual crop yields. This condition is prevalent in most tropical countries. The variations in the humidity and diurnal temperatures can project uneven conditions that make crops weak and so the hindered yields. The enhanced humidity levels can increase the foliage problems, root diseases and reduce yields. The crop outputs will take more time than expected due to increased humidity levels. The crop growth will be reduced and even damaged in most of the cases. The quality loss can be associated with the moisture and humidity levels. This will reduce the selling price and so most of the farmers can face economic losses. In order to enhance crop production rates, it is always essential to do proper monitoring. The wireless multi-sensors can be used to measure the moisture levels of the soils at different resolutions.

The optical crop sensors can be used to estimate the native crop conditions. These sensors work at different levels and at different resolutions. Most of the plants absorb red light and limited absorption is seen in the NIR range. The leaves that are in darker green can reflect more NIR radiation and can absorb red light. This is not the case with the leaves that are light green. The NIR and visible radiation can be used to develop vegetative indices that can help us in understanding the crop health. The Normalized Difference Vegetation Index or NDVI can be constructed using NIR and red wavelengths. The NDVI values range between -1 and +1. Most of the plants can give NDVI values between 0.1 and 0.9. The soil surfaces can yield an NDVI value of 0.1 to 0.2 whereas the canopies can yield 0.2 to 1. If the crop cover extends, then we can say that NDVI gets increased. Most of the sensors that are being used for crops are classified as active sensors.

The farming sector is encountered with several economic problems that affect costs and productivity. Shortage of labor can consume most of the income-generating activities. The concept of precision farming established certain standards that are needed for the agriculture field in current situations. Drones are available to the farmers at affordable costs and their usage has become prevalent in most of the communities. They can capture the images along with the GPS data. This feature in the drones can help us to know the characteristics of the croplands. RGB cameras and multispectral sensors can be installed in drones to identify the crop conditions. The ground sensor data and images derived from drones can be important in practicing precision agriculture in crop fields. Metrological situations can hinder the efficiency of the drones as they can alter the native utility of drones and sensing devices making the interpretation harder. The multispectral cameras onboard drones can be used to measure the vegetation patterns and monitor essential parameters. The parameters are NDVI, Leaf Area Index or LAI, ground cover and leaf water content. We can install thermal cameras in order to know the water stress conditions of the crops. Water stress can be due to enhanced temperate conditions. The stressed vegetation can be analyzed using thermal cameras. LIDAR systems and RGB cameras can be used to know the terrain surface. We can produce a Digital Surface Model or DSM and Digital Terrain Model or DTM of the area under study.

Internet of Things or IoT can be defined as Machine to Machine or M2M communications that can be implemented for the growth of digital markets and in this case agriculture. IoT can join technology, devices, process and people. The core idea of IoT is to associate humans with information technology. The technical abilities, collaboration and communications can be infused by virtue of this technology. All these components will give us data in real-time this enabling us in decision-making in a timely manner. This system has functional blocks. These functional blocks can facilitate the utilities like management, communication, actuation, identification and sensing. The IoT-based agricultural framework will include a physical layer that comprises of sensors, actuators, microcontrollers and network equipment. The network layer can comprise of internet, WiFi, GSM or CDMA and LTE. There will be a Middleware layer that can be called IoT Middleware. In the service layer, there will be an IoT cloud that comprises sensor acquisition, statistical analysis, crop disease information storage and equipment identification. In the IoT cloud, we can include livestock management, field crop management, cattle gaze service, pesticide control, message service, web service, image processing, video surveillance and expert service. In the analytics layer, we predictive analysis and multi-culture analysis. In the predictive analysis, there will be crop disease and detection of weeds, profit-loss analysis, data analysis, agro logistics, crop yield analysis and climate prediction. The multi-culture analysis includes aquaculture, floriculture, horticulture, olericulture, vermiculture, monoculture, arboriculture, silviculture and citriculture as components. The multi-culture analysis and predictive analysis can be combined and read as Big Data analytics. The user experience layer includes farmer experience that comprises dairy service, cold storage regulation, social network, farming community, fertilizer application, biomass power, biodiesel and resin extraction. IoT system can be dynamic in nature as it is self-adapting in nature. Most of the IoT devices are self-configuring in nature. This system has interoperable communication protocols. The devices used in this system have a unique identity like URI or IP address. It will be easier to integrate agriculture into core areas of information technology. This system has the self-awareness of context that can be offered by the data supplied by several platforms. With this technology, the decision making time will be relatively reduced. We can use IoT for an irrigation management system, control of disease and pests, monitoring of cattle movements, monitoring of dairy, monitoring water quantity and quality, monitoring and analyzing greenhouse conditions, soil monitoring, Precision agriculture and other applications. As far as the agricultural application of IoT is concerned, we can use automated hydroponics or Bitponics, Botanicalls, Edyn and Plantlink along with other equipment.

Water, food production and agriculture can be classified as unified commodities that are needed by humans at all times. 70% of water usage globally comes under agriculture (Hatfield, 2014). The spatial and temporal analysis of rainfall, temperature and others can be important. The elevation of the land can determine the local agrometeorological conditions. The STRV or spatiotemporal rainfall variability and drought in many places can be considered as important factors that can determine the fate of agricultural communities. The SRTV can reduce crop yields and this will produce several uncertainties in the farmer communities. This creates a lot of problems in the semiarid zones and contributes to decreased food security. Smart sensors can be used in case of precision agriculture and it monitors several factors that are required for the optimization of the crop growth levels. The environmental factors can be measured and analyzed with these sensors. With sensors in place, the farmers can know even microscale changes and hence can be prepared for environmental impacts and their reduction.

The crop yields can be sensitive to the local climatic patterns and weather anomalies. The increase in temperatures can enhance evapotranspiration rates and so the moisture levels of the soil will be affected. This can be considered as an important factor in the photosynthesis process. In the case of high temperatures, there will be enhanced phenological growth and this results in reduced growth periods. The yields will be reduced because of the heat stress conditions. The enhanced climatic variability can affect the phenological phases. The germination through the reproductive phase will be severely affected because of the heat stress conditions. In the case of maize, the climatic variability will reduce the photosynthetic rates and so this can be due to reduced leaf size. There will be lowered pollination if climatic variability is high and this will effectively reduce the size of the kernels. Climate variability affects 1/3rd of crops worldwide (Ray et al., 2015). This factor can be considered as important in the case of African countries. There will be negative crop yields. Climate variability can cause unfavorable conditions that can reduce crop yields. The sensors are classified into contact sensors and non-contact sensors. The contact sensors can be used to know the pressure, capacitance and can be shaft encoders and bubblers. The non-contact sensors are ultrasonic sensors, RADAR and MMC sensors. The pressure sensors can analyze hydrostatic pressure.

It is important that we use smart irrigation technologies for precision farming in developing nations to enhance yields. In the conventional irrigation systems or technology, there will be a fixed schedule irrespective of weather conditions and other factors whereas, in the smart irrigation system, we can analyze the climatic conditions and actual water content. The plant requirements can be considered important. In the existing irrigation system, most of the water will be wasted and with smart irrigation, we can save greater volumes of water with better precision. The plant productivity may not be based on effective irrigation. In the smart irrigation system, we can consider several aspects of plant growth patterns and needs. In conventional irrigation technology, automatic feed is not available whereas in sensor-based agriculture, this can be possible. We can use smart irrigation controllers that can provide the plant with accurate amounts of water at any time of the year. These smart irrigation controllers can be sensor-based or signal based controllers. Most of the sensor-based controllers can provide us real-time data pertaining to soil moisture, soil radiation, humidity, rainfall and temperature. We can use rain sensors that can get activated when there is a rainfall event. Many rain sensors are available in the market that can be used in monitoring irrigation systems.

Chapter

2

Sensors and Platforms

CONTENTS

2.1 INTRODUCTION

The sensors play an essential role in agricultural applications. The sensor applications in agriculture can be immense and it starts with land applications. This can be due to the coverage of the sensors at higher capacity if used on the terrestrial surfaces. The sensor applications in agriculture can differ and it is based on the needs of the farmer so as to produce good yields. The sensors can be used to know the economic features, land management and phenological variations. The precision agriculture methods need data based on the spatial scale. This data can be used to compare the crop fields pertaining to crop response. This data can be useful to know several biophysical variables that can dictate the crop growth patterns. This data can be used in managing the crops.

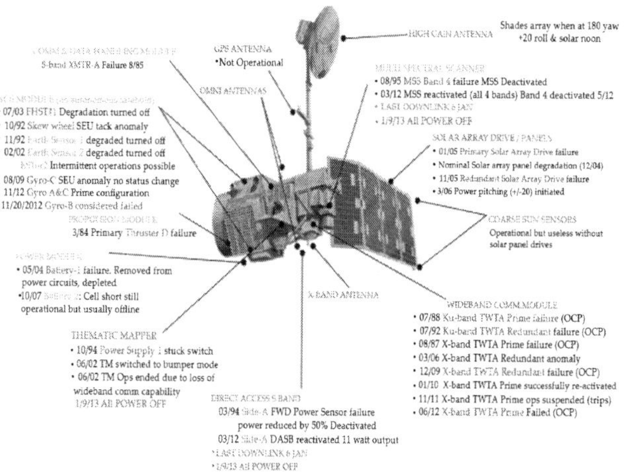

Figure 2.1: Landsat TM.

Source: https://eos.com/landsat-5-tm/

The satellite data and crop yield monitors can be used to collect the data that is needed for precision agriculture. Data for the entire growing season may not be possible and this can be due to several operating factors. The dense sampling can be always challenging. Crop modeling and remote sensing can be considered as an important alternative to the field-based sensors where feasibility is less. The remote sensing technique has emerged as an important tool in the estimation of biomass and several other parameters (Price et al., 1991).

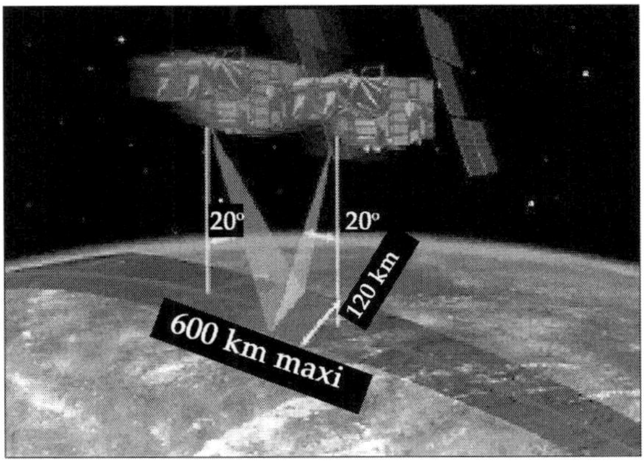

Figure 2.2: SPOT.

Source: https://eoportal.org/web/eoportal/satellite-missions/s/spot-5

We can easily perform soil mapping and stress mapping using this technique. Popular sensors are Landsat TM (Figure 2.1), SPOT (Figure 2.2) and aircraft-based sensors. Most of the data provided by the sensors free of cost suffer from low resolution. It is important to go for high-resolution data in order to demarcate essential crop patterns and manipulations that are due. The cost of this data is much higher and may not be available to the farmers.

Most of the vegetative characteristics of the crops can be analyzed by remotely sensed imagery and it has become an essential need these days. Most of the vegetation indices are based on plant physiology and they can be derived from the images. The vegetation indices can be framed according to the red and near IR radiation. The physical properties of the leaves can be observed using the absorbance at various wavelengths. The time-series images can be important in interpreting the changes that occur in the growing season. The vegetation indices that were computed can help us to understand the spatial and temporal patterns that determine the fate of the crops. These vegetation indices may not be associated with the biophysical parameters always. Polynomial and linear regressions can be made through the analysis of images collected from two dates. They can be correlated with the field measurements like plant height, biomass, leaf area index and yield. We can use the images post-processing to compute transformed soil adjusted vegetation index or TSAVI, soil adjusted vegetation index or SAVI, simple ratio index or SR and normalized difference vegetation index or NDVI. The TSAVI can be calculated using the following equation:

TSDAVI = 1.20 X (NIR-1.20 X Red-0.08) / 1.20 X NIR + Red – (1.20 X 0.08) + X (1+1.20 X1.20) and SR can be calculated as NIR/Red. SAVI can be calculated by NIR-Red/NIR +Red +L X (1+L).

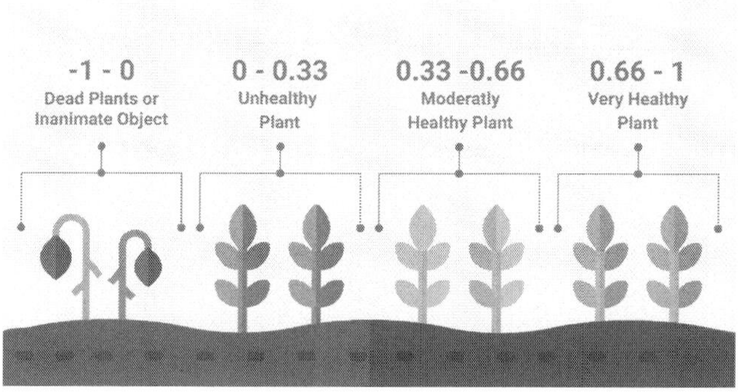

Figure 2.3: NDVI.

Source: https://eos.com/blog/ndvi-faq-all-you-need-to-know-about-ndvi/

NDVI (Figure 2.3) can be calculated by NIR-Red/NIR + Red. The vegetation indices calculated using remotely sensed images can be used to estimate the growth variability and health conditions. Using spectral reflectance of the crops, we can easily quantify the vegetation growth kinetics. NDVI is being used in several monitoring applications pertaining to crops. We can use a single image to know the exact yield and others. The phenology estimates can be obtained using remote sensing images. In order to understand crop phenology, it is important to have a high temporal resolution. The phenological variations can be studied with MODIS or Moderate Resolution Imaging Spectroradiometer images. Most of these sensors can measure the radiation emitted and reflected from the object under study. Light is a form of electromagnetic radiation that reaches the earth's surface with varying wavelengths. The electromagnetic spectrum comprises of both shorter wavelengths and longer wavelengths. The shorter wavelengths are gamma radiations and X-rays. The longer radiation comprises of radio waves and microwaves. Human eyes can see only visible radiation. The spectral wavelengths can be called bands. The chlorophyll content of the leaves can be used as a standard to estimate the crop health kinetics. The reflection of light can be associated with the concentration of the pigments that are present in the leaves. The leaves that are having high chlorophyll content can be green in color. Due to the presence of green content, the NIR reflectance will be higher. The water-stressed leaves can exhibit low rate of reflectance. The estimation of variations in the reflected light wavelengths can help us demarcate vegetation species, vegetation, green content or chlorophyll and soil vegetation. The remote sensors can be produced with various grades of sensitivity. The resolutions provided by the remote sensing instruments can be classified as spectral resolution, spatial resolution, temporal resolution and radiometric resolution. Based on these resolutions, we can classify sensors.

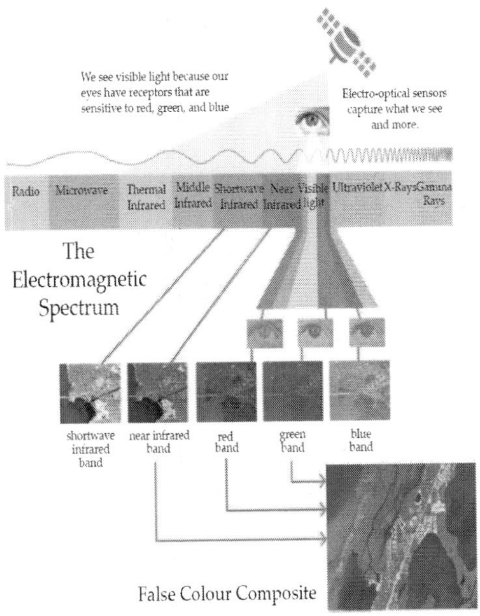

Figure 2.4: Spectral resolution.

Source: https://www.slideshare.net/Dhwani7887/remote-sensing-66205597

Spectral resolution (Figure 2.4) can be defined as the sensitivity of sensors to different wavelengths in the electromagnetic spectrum. The sensors can be classified into hyperspectral and multispectral sensors. The hyperspectral sensors can be sensitive to 100 narrow wavelengths and multispectral sensors can be sensitive to 10 broad wavelengths. The spatial resolution can be defined as the sensor pixel size. Pixel can be considered as an indivisible feature that a sensor can record. The size of the pixel can be based on the distance between the target and sensor. The viewing angle of the sensor can be important. Instantaneous field of view or IFOV can be considered as an essential factor that is to be considered whilst understanding the spatial resolution. IFOV can be an area that is visible via a sensor at any point in time. In the case of low-resolution images, we can observe large features and phenomena easily. In the case of high-resolution images, we can identify both large and small objects. The low-resolution images will have a large pixel size whereas high-resolution images will have a small pixel size. The images captured by QuickBird can be of high-resolution images that can give us 2.4 m pixel size. This resolution can be important to know the crop growth kinetics. The thermal infrared imagery may be of low resolution with 5m pixel size. It is always important to use high-resolution imagery to know the crop growth kinetics accurately. It is important to use 0.5 m pixel size imagery is available for a better understanding of the growth patterns and kinetics.

Temporal resolution can be defined as a revisiting period of a sensor onboard satellite. Some sensors can give us data every day and some can give it twice a month. It is important to know that the characteristics of any object or phenomena can change with time and so the temporal component is considered as important and it can be studied as temporal resolution. The change in the crop growth and yields can be detected with the sensors with high temporal resolution and that means the frequency of the satellite to view a point repeatedly. Weather can infect the quality of the data provided by the sensors. The radiometric resolution can be defined as an ability of a sensor to demarcate the energy intensities in the electromagnetic spectrum or energy. If the sensor can detect several differences in the energy intensities, we can classify that sensor as having a high radiometric resolution. The image brightness can be based on a number of bits that can be utilized to display energy sensed by the sensor. In general, the brightness values of a black and white image can be represented in a range starting from 0 to 255. 0 being white and 255 is black. This can be called an 8-bit image.

The remote sensors can be classified based on the platform they were installed. They can be classified based on the spectral, spatial and temporal resolution along with radiometric resolution. The light source can be taken as an important factor in the classification of the sensors. Most of the platforms on which the sensors are installed are owned by the governments. Some of the sensors can be installed on all-terrain vehicles, agricultural machinery, kites, balloons, unmanned air vehicles, and airplanes. The platform on which these sensors are placed or installed can determine the quality of the data. The sensors can be classified as active and passive sensors. The active sensors require energy from the external source whereas the passive sensors may not need the energy source from an external source.

The passive sensors can be used only if there is sunlight that exposes the target area. The cloud cover must be minimal so that the sensors can acquire images correctly. Due to cloud cover, the passive remote sensors are restricted to noon times only. These passive sensors are mounted on QuickBird and Landsat or on the airplanes. The active sensors need energy from the external source and they can be operated at any time. We can fix the wavelength of these sensors. These sensors can illuminate the objects and can absorb light reflected back through photodiodes and can measure it. Active sensors were mounted on ground vehicles and satellites. The active sensors like CropCircle and GreenSeeker can be mounted on the vehicles and Radarsat on satellites. Remote-sensing technology can offer us several uses. It can be used in drought monitoring, land use mapping, land cover classification, environmental monitoring and agronomic management. The remote sensors can be to detect even the minimal variations in the vegetation growth patterns. We can use these remote sensing tools to quantify crop yields. We can calculate the vegetation indices using these sensors. The reflectance can be read in the form of spectra bands. The vegetation indices can be used in evaluating the vegetation growth kinetics. The colors that are present in the images derived from sensors can be used to interpret crop growth.

We can classify these areas and related them with the biomass concentrations. The biomass differences can help us quantify crop growth effectively. There are several other vegetation characteristics and attributes that canopy exhibit. These characteristics can be handy while setting the band combinations of the sensors. Most of the vegetation indices can help us know water stress, leaf area index, biomass and chlorophyll content. The variations in the vegetation conditions can be represented by the VI values. Most of the changes can be represented by the variations in the spectral reflectance. The vegetations that are green can absorb visible light and can reflect NIR radiation. If the vegetation understudy is stressed it can reflect most of the visible light and the least amounts of NIR radiation will be reflected. If the NDVI levels are less, then we can interpret that the vegetation in that area is chlorotic and so low biomass. If NDVI values are high, then we can assume that the plants are healthy with high biomass. We can use remote sensing tools in nutrient management practices. We can detect the nitrogen deficiency symptoms via variations in the spectral reflectance. We can use NIR and red portion of the electromagnetic radiation. We need to ensure that at least 1-meter resolution is to be maintained. It is important to use NIR and Visible wavelengths to achieve better results. The temporal resolution must be at least 15 days. We can use images derived from Rapideye and GreenSeeker and CropCircle. We can use NDVI and GNDVI for this purpose.

Using remote sensing technologies, we can reduce nitrogen losses. We can even apply nitrogen at specific sites. We can use these techniques in irrigation management. We can assess the water stress conditions by interpreting the thermal images. The resolution in this can must be at least 1 meter. We can use NIR and visible radiation for this purpose. The temporal resolution can be 7 days or even 1 day. We can use cameras that can support IR and can capable of extracting images in NIR and mid-IR images. We can use NDWI and WI for this purpose. We can go for irrigation scheduling and variable rate irrigation using this technology. We can predict yields effectively. The spectral differences sensed by the remote sensing tools can help us to predict the yields. The minimal resolution possible can be 1 meter. We can use visible and IR wavelengths for this purpose. We need to maintain that the observations be made twice in a growing season. We can use WDRVI, GNDVI and NDVI for this purpose. We can go for site-specific management using this technology. We can manage pests and several other problems associated with plants. The stress caused due to diseases can be estimated using remote sensing tools. The plants can emit heat and this can be sensed by the sensors. We can get the thermal images. The spectral variations can be identified using NIR and VIS wavelengths. The temporal resolution can be 7 days. We can use both active and passive sensors. Through this technology, we can assess the need for fungicides and pesticides. Soil mapping can be done using these techniques. We can interpret the images based on the differences in the soil colors. We can use NIR and Red portions of the electromagnetic spectrum to achieve better results. The spatial resolution must be at least 1 meter. We can use NIR and visible wavelengths. The temporal resolution must be at least 7 days. We can use both active and passive sensors. These techniques can be used to know the variable rates of nutrient application and soil sampling.

2.2 REMOTE SENSING

Remote sensing can be defined as a science that help us acquire information about the surface of the earth without any physical contact. We can sense and record the energy reflected through remote sensing devices. There will be certain interactions between the radiation and the target in question. We can use image systems for this purpose. We can use some sensors that can be classified as non-imaging sensors. The energy source can be called an illumination source. This can be called an important and primary requirement of any remote sensing equipment. The energy can reach the target through the atmosphere. Due to its interaction with the atmosphere, the energy that is incident on the target may suffer from transmission defects. The energy interactions with the target can be based on the target properties. The incident energy can be scattered and can get emitted by the targets in question. It is important to establish a sensor to detect the electromagnetic radiation reflected or emitted from the targets. Transmission followed by reception and processing can be considered as important steps in any remote sensing operation. (Figure 2.5)

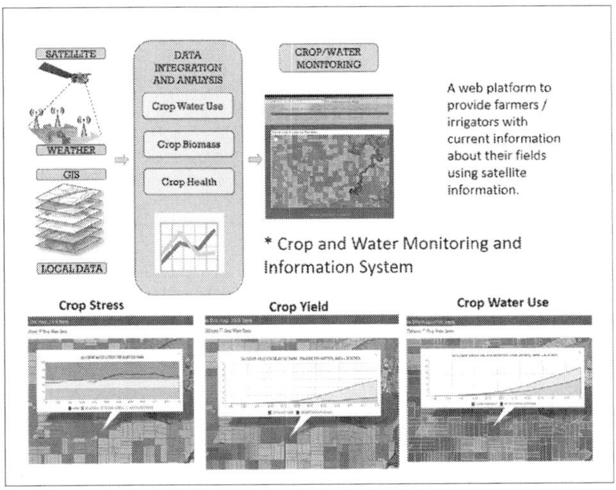

Figure 2.5: Remote sensing.

Source: https://uwrl.usu.edu/water-resources/projects/remote-sensing

The data captured by the sensors will be stored in either digital format or as a hardcopy. The image capture will be processed. There will be an interpretation of data to extract the information needed. The energy will be in the form of electromagnetic radiation. Most of the electromagnetic radiation behave in ways that are predictable and it was supported by the wave theory. The EMR or electromagnetic radiation comprises of an electrical field and magnetic field. The electrical field can be perpendicular to the magnetic field. It can be stated that the electric field propagates at right angles against the electric fields. The electrical field and magnetic field can travel at the speed of light. The wavelength and frequency can be considered as important characteristics in the entire process of remote sensing. The wavelength can be defined as the length of a wave cycle. In general, this can be calculated by taking into consideration the distance between the crests. This can be measured in meters and other units. The frequency can be defined as the number of cycles that a wave possesses while traveling a point. We can calculate frequency in Hertz. The wavelength and frequency are inversely proportional. If the wavelength is short, then we can assume that frequency is high. The electromagnetic spectrum can extend from shorter wavelengths to longer wavelengths. The shorter wavelengths are gamma and x-rays mostly. The longer wavelengths are microwaves and radio waves that can broadcast. We can use most of the electromagnetic spectrum in the remote sensing process. The ultraviolet portion of the spectrum can be classified as having a short wavelength and hence greater frequency. Some of the minerals, rocks can fluoresce the UV radiation. The light which we can sense comes under the visible spectrum. The visible spectrum can be considered as relatively short. Most of the radiation cannot be experienced by us. The remote sensing instruments can detect several types of radiation around us. Red has the longest wavelength in the visible spectrum and violet has the shortest. The concept of colors can be associated with the visible spectrum only. The violet region can extend from 0.4 micrometers to 0.446 micrometers. The red region can extend from 0.62 to 0.7 micrometers. The blue, green and red colors can be classified as primary colors. The infrared region can be called as IR region and it extends from 0.7 to 100 micrometers. This region can be divided into reflected IR and thermal IR. The thermal IR can be called emitted IR. The reflected IR extends from 0.7 to 3 micrometers. The thermal IR can extend from 3 to 100 micrometers. The heat emitted from the earth's surface can fall in the thermal IR range. The microwave region of the spectrum can extend from 1mm to 1m. This region comprises of longer wavelengths. The short wavelengths in the microwave region share the same properties as the thermal IR region. The longer microwave wavelengths can reach the wavelength of the radio broadcasts. The radiation that reaches the earth's surface should interact with the atmosphere. The gases and particles that are present can alter the properties of the spectrum. There will be absorption and scattering of radiation. If the incoming solar radiation gets deviated from its native path due to the presence of large gas molecules and particles, then we can call that process scattering. Scattering can be based on many factors. Some of the factors are distance through which the radiation passes in the atmosphere, presence of gases and particles and wavelength of radiation. The scattering process can be classified into three types.

Rayleigh scattering (Figure 2.6) generally takes place if scattering occurs due to particles that are smaller than the radiation wavelength. Dust specks, oxygen and nitrogen gas molecules can bring about Rayleigh scattering. Due to this scattering, the energy with shorter wavelengths can be shifted to longer wavelengths. This type of scattering is common in the upper layers of the atmosphere. The blue color of the sky is due to this type of scattering. The scattering occurs only if the particle size is equal to the radiation wavelength. Water vapor, smoke, pollen and dust can bring about Mie scattering. This scattering affects longer wavelengths. This occurs in the lower portion of the atmosphere. The presence of larger particles is evident in the lower atmosphere. The nonselective scattering occurs if the particle size is larger than the radiation wavelength. The large dust particles and water droplets can bring about this scattering. This scattering can occur if there are fog and clouds in the lower atmosphere. Nonselective scattering is the reason why clouds appear white.

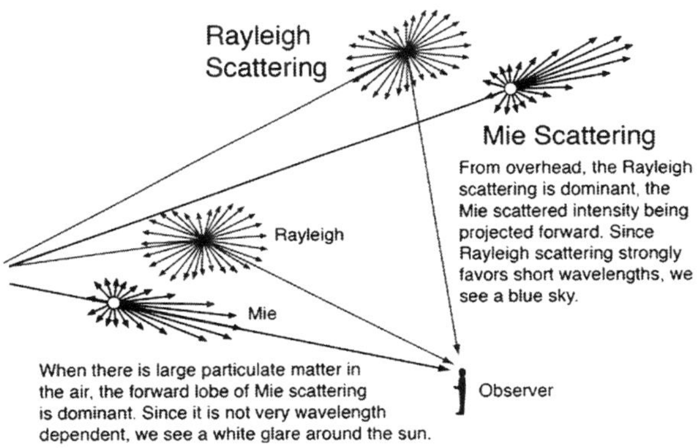

Figure 2.6: Rayleigh scattering.

Source: http://hyperphysics.phy-astr.gsu.edu/hbase/atmos/blusky.html

Electromagnetic radiation can interact with the atmosphere and gets absorbed. This process can be called as absorption. Different molecules that are present in the atmosphere can absorb energy. Water vapor, carbon dioxide and ozone can absorb most of the radiation at diverse wavelengths. The ozone layer can absorb most of the ultraviolet radiation that will be emitted from the sun. Carbon dioxide can be classified as one of the greenhouse gases. This gas can absorb far-infrared radiation. The heat generated by this radiation will get absorbed by the atmosphere. Water vapor can absorb longwave infrared. It also absorbs microwave radiation. The gases that are present in the atmosphere can absorb radiation in specific locales. We will use those regions present in the electromagnetic spectrum that are not affected by the atmosphere and hence can be used in the operation of remote sensing systems. We call these areas that are least affected by the atmosphere can be called as atmospheric windows. (Figure 2.7)

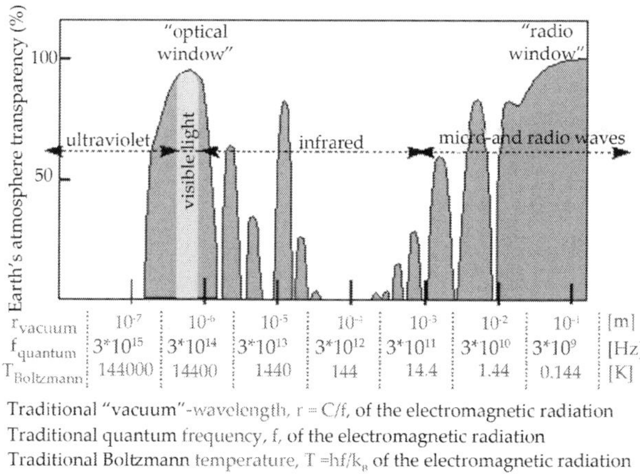

Figure 2.7: Atmospheric windows.

Source: https://www.gb.nrao.edu/GBTopsdocs/primer/atmospheric__windows_.htm

The radiation that is neither scattered nor absorbed by the gases or any particles in the atmosphere can reach the earth's surface. There will be interactions between the objects present on the surface of the earth and radiation. There will be three modes of interaction. They are classified as absorption, transmission and reflection. Absorption is a process in which the radiation is absorbed by the target. Transmission is that if the radiation passes via target. Reflection is that if the radiation bounces off the target of our interest. In the remote sensing technique, we need to measure the amount of radiation reflected from the physical objects. The energy can be reflected in two ways. They are diffuse reflection and specular reflection. The specular reflection can be observed if the surface is smooth. This type of reflection can be called as mirror-like reflection. The light that is incident on the surface is reflected in one direction only. In the case of a rough surface, we can find diffuse reflection. The incident radiation can get reflected in all directions uniformly. The physical objects on the surface of the earth can act as diffuse reflectors and specular reflectors and in between them. This can be based on the roughness of the objects. If the particle size and surface variations are larger than the incident wavelengths, then we can expect diffuse reflection. The grained sand can appear rough when sensed through visible wavelengths and will appear smooth if sensed via microwaves.

Leaves have chlorophyll and so they absorb blue and red wavelengths and the green wavelengths will be reflected. The chlorophyll content in the leaves can be high in the summer season and less chlorophyll content is seen in autumn. If the leaves are healthy, they can act as diffuse reflectors of NIR radiation. Water absorbs longer wavelengths like visible and NIR radiation. The algal members have chlorophyll and they can absorb blue wavelengths. They can reflect green wavelengths. Based on the reflected and emitted radiation levels of the targets, we can construct spectral libraries that explain spectral response. The response patterns obtained by the remote sensing equipment can be compared with the standard spectral libraries and then the images can be interpreted accordingly. The spectral responses can be varying according to the season and weather conditions. The electromagnetic energy can be detected electronically or photographically. We can use ground-based sensors to collect information about the physical objects on the surface of the earth.

The path taken by a satellite can be called as an 'orbit'. The satellites that are placed at high altitudes can be called geostationary obits and they can view a point on earth continuously. These satellites are placed at a height of 36000 kilometers. They revolve at the same speeds as earth and so can stay at one point only. These satellites can be used to monitor cloud patterns and weather conditions. Most of these satellites can follow North-South directions and can cover most of the earth's surface. These orbits are called near-polar orbits. The satellites that follow these orbits can cover the day time of the target area. These orbits among near-polar orbits can be called sun-synchronous orbits. These orbits allow the satellites to follow local sun time or local day time. Most of the satellites can be placed in near-polar orbits. These satellites can travel towards the southern pole on the first half of the earth and towards the north pole on the second half of the earth. These satellites can be called as descending and ascending passes. The sensors can sense certain areas on their path and this path can be called a swath. Agriculture can be classified as an important commodity for any country. The conventional methods of information capture can be time -and tedious factors. The remote sensing equipment can give us information regarding the cropping area and other attributes. Remote sensing technology has changed our understanding of several environmental phenomena. Geographic Information Systems with remote sensing can be used in natural resource management initiatives. The spectral response can be affected by several plant parameters. Some of the parameters are internal leaf morphology, leaf architecture, nutritional status and pigmentation. The leaves can exhibit specular and diffuse reflections. The diffused reflection from the leaves can be attributed to the internal leaf morphology. Inside the leaf, there will be multiple scattering. The specular reflection can be attributed to the leaf surface. This can be linked with the leaf hairs and cuticular waxes. The spectral reflectance can be used in crop inventory. The data generated from the spectral reflectance can be used to demarcate crops and estimate crop growth. We can assess the condition of the crops that might be suffering from nutritional factors and others. Every crop has its own spectral signature and this can be due to the internal and external morphology of the plants. The growth conditions of the crops can remain an important factor in exhibiting spectral reflectance.

2.3 DATA ANALYSIS

The ranchers and farmers can observe the fields and other parameters without coming into contact with the plants using remote sensing and GIS. We can know the health condition of the crops by observing the color of the leaves. We can assess the field conditions without physical contact using sensors onboard satellites and aircraft. Most of the sensors are sensitive to visible radiation. Remote sensing techniques can provide data in the form of images and we can develop agricultural statistics from that. The airborne instruments and satellite sensors can be considered as important sources. The airborne instruments can provide large-scale ortho-imagery that is spatially obvious and we can easily demarcate the features present on the surface of the earth. These instruments can be used in the surveying process. We can use hyperspectral sensors for this purpose so that spectral signatures can be easily analyzed. Synthetic Aperture Radar or SAR and LIDAR techniques can be used in the surveying process. (Figure 2.8,2.9)

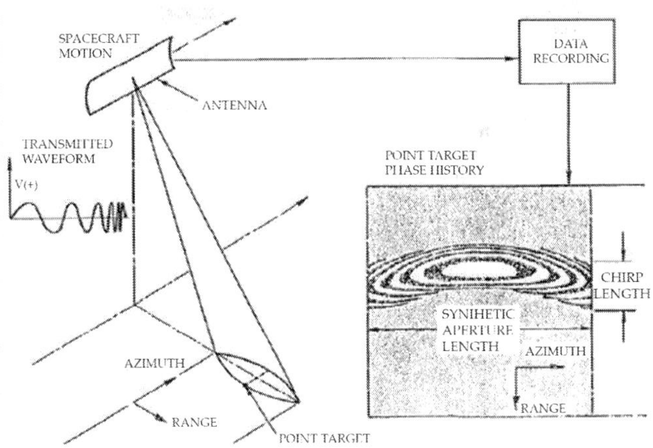

Figure 2.8: Synthetic Aperture Radar.

Source: https://www.usgs.gov/media/images/principles-synthetic-aperture-radar

Figure 2.9: LIDAR data.

Source: http://calag.ucanr.edu/Archive/?article=ca.v069n01p14

Figure 2.10: RISAT.

Source: https://earth.esa.int/web/eoportal/satellite-missions/r/risat-1

Synthetic Aperture Radar can provide us high-resolution images at a resolution of 4 to 30 meters. RISAT (Figure 2.10), ALOS-PALSAR-2 (Figure 2.11) and Radarsat-2 (Figure 2.12) can provide high-resolution data. The temporal resolution of these satellites is 24 and 6 days respectively. We can use these images to delineate crops and their identification. The resolution will be 20 meters. It is possible to extract images in day and night times. We can use MODIS and AVHRR data for the assessment of crops. Based on the type of sensor and processing levels, the images may need processing. The satellite images can satisfy almost all the technical requirements. It is important that we register different bands in the case of multiband data.

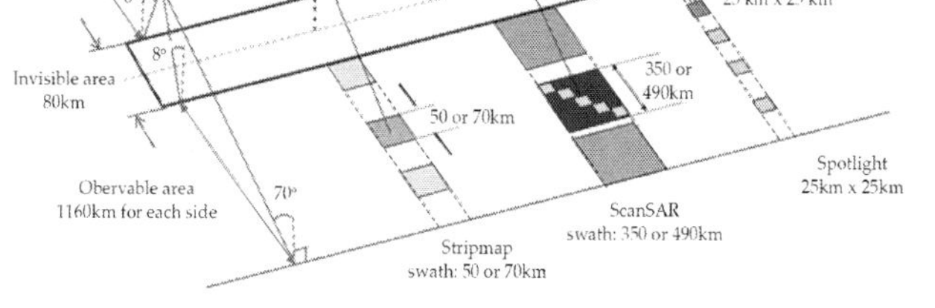

Figure 2.11: ALOS-PALSAR-2.

Source: https://www.eorc.jaxa.jp/ALOS-2/en/about/palsar2.htm

The image data obtained should be tagged with the metadata. The metadata is important because it will provide all the information on the sensor platform. The metadata can give us information about the position and time of the acquisition, solar illumination and other calibration parameters. The data pertaining to the cloud cover and missing information will be provided by metadata. The suppliers will detect and rectify data as and when needed so that the end-users can use the data easily. Atmospheric correction and ortho correction can be classified as important steps in the processing of multispectral images. There will be terrain correction and sensor-specific correction in the ortho correction process. In case of terrain correction, we need to use the digital elevation model or DEM so that the georeferencing process can be made with the least effort necessary. In order to correct the angles of the images, it is important that we go for terrain correction.

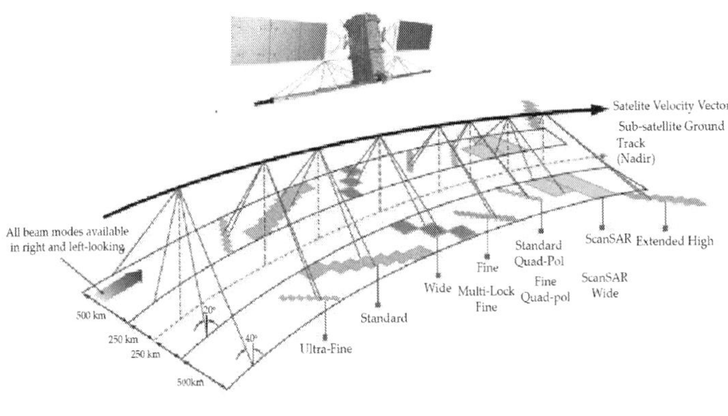

Figure 2.12: Radarsat-2.

Source: https://eoportal.org/web/eoportal/satellite-missions/r/radarsat-2

Atmospheric correction can rectify the atmospheric effects. It deals with BOA or Bottom of Atmosphere and TOA or Top of Atmosphere parameters. BOA and TOA can be important in image processing. The BOA can be equivalent to Top of Canopy or TOC that can provide us the reflectance values of the pixels. These values can build raster imagery. The multispectral bands and panchromatic bands can be merged to get high-resolution images. In the case of SAR images, there will be calibration and geocoding. In the calibration process, the band intensity of the SAR image can be converted into backscattering coefficients. The SAR can be side-looking and its specificity can be based on range and azimuth directions. Geocoding can be done to compensate for these effects. It can be rectified by using filters like speckle filtering and multilooking filters. Imagery is classified into raw data or 1A and system corrected 1A levels or 1B. 1C level can be called a georeferenced level. 2A level can be classified as thecalibrated level. The 2A level can be supplied with certain fees. Some of the software can be used to calibrate this data. The single scenes can be processed via mosaicking so that large coverages can be possible. There will be segmentation, classification and time composting. It is essential to use appropriate software so that there will be the least error propagation. There are several open-source software's available in the market. We can use C++, Java and Python libraries to handle images. Popular remote sensing toolkits are Orfeo ToolBox and Sentinel specific toolboxes. We can use QGIS and GRASS so that we can perform geospatial visualization and analysis. Some of the raster database solutions are PostgreSQL and Postgis. The spatial web server solutions are Mapserver and Geoserver. Most of these operations can be performed by GDAL or Geospatial Data Abstraction Library. Some of the commercial software are ESRI-ArcGIS, PCI, ERDAS Imagine and Excelis ENVI/IDL. In the case of Sentinel satellites, we get the SNAP tool or Sentinel Application Platform that can help us in understanding several functionalities.

Chapter 3

Mapping of Crops

CONTENTS

3.1 INTRODUCTION

The knowledge of trends in agricultural yields at a national scale can be important in order to face the threats that may arise in the future and present. These threats can jeopardize reliable access to food and its stable supply. The fluctuating meteorological conditions and management decisions can make our understanding of crop yields complex (Fischer et al., 2005). The forecasting of the food supplies can give us an estimate of the crops that are sown. It is always important to use mapping techniques to know the efficiency of the cropping patterns. It should be kept in mind that the data derived from remote sensing devices are costly. These images can be useful in extracting the exact information about the crops. The cropping systems can be classified as diverse and in most cases, we know that they are complex. The success rate of the remote sensing products can be based on their adaptability to the local cropping systems with native environmental conditions. The classification of the satellite images can be considered as an important factor in crop mapping protocols. We can use automated techniques to map crop yields. The surface types of the crop fields can be monitored effectively by remote sensing devices. The image data can be stored in the target classes and this will be done by using spectral characteristics. Most of the crop maps can be produced using a process in remote sensing called supervised classification (Beltran et al., 2020). In supervised classification, human intervention is necessary.

In the case of supervised classification, an analyst will use the ground observations to recognize the location in the images and relate them to the target classes that can be mapped. These target classes are obtained with the aid of training samples. With the help of training classes, we can train the computer to know the features that are present in the satellite imagery. Then, the analyst uses classification algorithms to match the image data with the reference points. In some cases, we can use secondary data also. This classification algorithm can statistically define the characteristics of the classes. It compares the related pixels and associate them with the target classes. The accuracy of the classification can be measured. This called the validation step. We use the ML or Maximum Likelihood classification method in the case of supervised classification of images (Lu et al., 2015). In this type of classification, we use the density function of multivariate probability with the aid of training data to define classes based on statistics. These density functions represent the probability of the spectral pattern of every class (Denegre, 1994). The data can be associated with the training classes that can have a high rate of likelihood (Jensen, 2019). This approach is widely accepted in agricultural mapping techniques (Rogan et al., 2007).

In the process of supervised classification (Figure 3.1), we consider in situ data or field data, EO data or satellite images and secondary data. The field data must be checked for consistency and this step is called quality control.

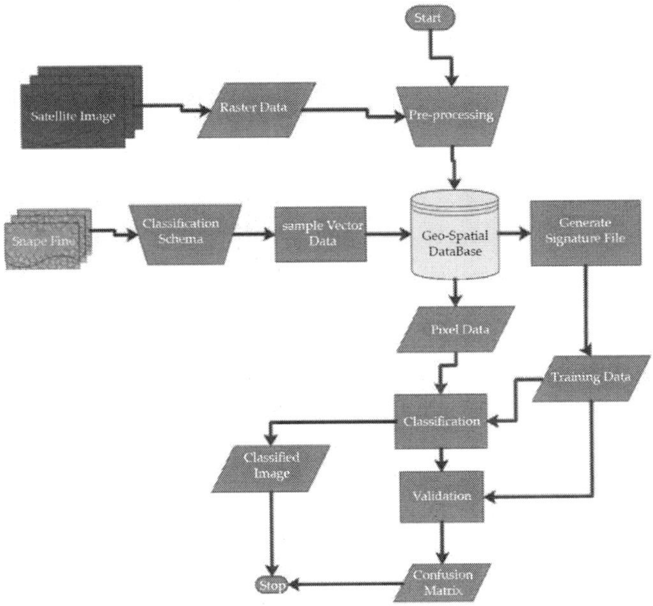

Figure 3.1: Supervised classification.

Source: https://www.researchgate.net/figure/Flow-chart-for-Supervised-classification-of-satellite-image_fig1_306304077

The EO data should undergo pre-processing and the secondary data must provide information that is relevant in nature. Post quality control, we need to split data and this process is called data splitting. The data can be split into the validation sample and training sample. In the pre-processing stage, we can perform geometric correction, radiometric correction, atmospheric correction and clouds screening processes. In the case of secondary data, we need to extract information like geology, aspect, slope and elevation. After the image pre-processing stage, there will be a spectral band selection step. In this step, we will perform pixel-based classification and image segmentation or feature extraction. The image segmentation and feature extraction will be based on the imagery and secondary data. This will result in the object-oriented classification. In this process, we use support vector machines, random forest, decision trees etc. We need to perform parcel delineation followed by post-classification processing. In this process, there will be mosaicking and filtering, and then accuracy assessment. The training samples can be classified using the classification techniques like maximum likelihood, random forests ,etc. In the accuracy assessment step, we will use validation samples in order to get user accuracy and producer accuracy. Map classification will be done. Post this step, we can get value-added products.

The supervised classification of the images is being used for decades and this will be done pixel/pixel basis. The spatial variability of each band will be less within each crop. This can be high in the case of different crops. Research is focused on generating classification algorithms that can demarcate homogenous pixels. These aggregations can be called objects. We can produce or generate the image objects and this process can be called image segmentation (Peter et al., 2011). The main aim of this process is to generate objects that are spectrally and spatially heterogenous (Comaniciu et al., 2002). The features can be extracted using attributes, morphic attributes, textual and spectral characteristics of the objects (Blaschke, 2000). The image objects can be subjected to classifications and this can be called as object-oriented classification. This technique can be used in the identification and mapping of the crops (Negri et al., 2011). In the classification process, we can use multiple data types. They can be ground data, satellite data and ancillary data. If the data obtained is accurate, then the products derived from them can be accurate (Thenkabail et al., 2008).

The quality of ground data can be of utmost importance in mapping croplands. This is due to the fact that the ground data are used to train in the supervised classification and this data can be used in the validation of maps. The accuracy of the field or ground data can determine the accuracy of the classification process. It is essential to collect information from the field as accurately as possible. The classification of the crops can be done through field surveys. It is important to classify crops based on field surveys. We can record growth stage, cultivar, crop type and GPS points via field surveying. The spatial and temporal values of the samples can influence the accuracy of our interpretation of the images during the cropping season and then mapped according to the requirements. This technique can be important to know the phases in the crop cycle and they are the reproductive phase, fruit and flower development phase. We can limit our observation to one/field/season. It is hard to know the size of the sample that is needed to train the classifiers. The accuracy of the classification get increased if the number of fields are increased to train the classifiers. The ground sampling process can result in oversampling or undersampling of the crops. The main problems with the field surveys is that they are challenging, laborious and time-consuming. The main aim of any agricultural field survey will be to attain a representative sample at regional and national levels. The earth observation data or EO data can be important in monitoring agricultural activities (Fritz et al., 2019). A minimum number of images needed and time of acquisition is important and this data is specific to the region under study. We can use synthetic aperture radar or SAR, but it has certain limitations. The dynamic nature of agro-ecosystems can be considered as an important factor. The geographical extent that is required to be mapped can affect the acquisitions. Data availability, engineering constraints, spectral similarities, diverse cropping patterns, diverse cropping cycles and minimum mapping units can be considered as important factors in deciding the remote sensing systems that are to be used (Waldner et al., 2018).

Figure 3.2: Global Agricultural monitoring (GEOGLAM).

Source: https://glam.umd.edu/project/geo-global-agricultural-monitoring-geo-glam-initiativethe-geo-global-agricultural-monitoring

Global Agricultural monitoring is an initiative that focuses on agricultural monitoring. It is also called as GEOGLAM. Some of the target products of GEOGLAM (Figure 3.2) are crop mask, cropping systems, environmental variables, biophysical variables, crop yield and crop acreage.

The optical remote sensors can collect information in Visible, NIR and SWIR regions. These sensors can be useful in monitoring vegetation and we can easily demarcate the spectral signatures of different cropping areas. Agricultural mapping can demand multispectral acquisitions. The minimal resolution required must be 10 meters. This can be achieved from Landsat images and data is free of charge (Fisette et al., 2014). At this spatial resolution, we can demarcate the crop fields as individual entities. We can use AWIFS data, SPOT and MODIS data to fill the gaps. These gaps can be due to different revisiting times and cloud cover. Synthetic Aperture Radar sensors can be operated at microwave frequencies. We can calculate the intensity of energy emitted back after target interactions. The phase of energy can be important in these operations. We can get data at all times as these sensors have their own energy source. The microwaves used in SARs have low frequency and so they remain unaffected by the atmosphere. We can get data from the regions where cloud cover is high. The synthetic aperture radar data can be used to demarcate vegetation types and this can be due to scattering of long wavelengths. This scattering can be crop-specific. The scattering can be attributed to the presence of fruits, stems, leaves, shape and size. We can extract the volume of water that is present in the canopy. This can be detected at a molecular level also. The roughness and moisture of the soil can affect the radar signal. The configuration of the sensors, growth stage, crop type and other crop properties can affect the soil conditions. Most of the SAR sensors can be based on polarization, incident angle and frequency. The frequency of the sensor can define the depth of penetration into the canopy. This signal intensity can be altered by canopy entities like stalks, fruits and leaves. The microwaves can interact with the dominant plant parts. It is important to define the optimal SAR polarizations. We need to use linear cross-polarization to separate the crops. This can be HV or horizontal vertical or VH or vertical-horizontal polarization (Sonobe et al., 2017). There will be repolarization of the signal due to multiple scattering. We can use VV or vertical-vertical polarization to identify the crops. The vertically transmitted waves can help us to know the vertical structure of the crops. Out of all these polarizations, HH or horizontal-horizontal polarization can be considered as least important as the information generated will be less obvious.

Ancillary data can be called secondary data and it can be obtained from certain sources that may not be considered as remote sensing equipment. The ancillary data can be used to enhance the accuracy of the classification system. Ancillary data refers to additional variables that can be linked with the data obtained from the remote sensing devices. All the aspects of ancillary data will be used in the classification process. Some of the variables are vegetation maps, political boundaries, transportation networks, soils, geology, hydrology, aspect, slope and elevation that can be used in enhancing the accuracy. Most of the agricultural monitoring systems need to provide standardized information with extreme precision and accuracy at all time. The methodologies should be robust and can be repeated (Franklin et al., 2010). There are some problems that can hinder our usage capability of this data and this can be due to several reasons. Some of them are lack of scientific understanding and hurdles in the operational utility. The operational systems that can be used in extracting information from the agricultural fields can be called dynamic data and it is based on the in situ data networks.

Some of the global monitoring systems are USAID, FAO, MARS and IRSA (Becker -Reshef et al., 1994). If we need data to monitor agriculture on a national scale, we can use 30m resolution data. The ground truth data can be used in the training of classifiers. The misidentification of crops is quite possible and this can be due to operation standards set whilst those acquisitions.

3.2 ESTIMATION OF CROP AREA

The information about crop production can be considered important in planning agricultural activities. In any crop information system, there will be two important components. They are called as crop yield and crop area. The estimation of crop yield is a complex process that area estimation. The estimation of crop area can be complex sometimes. The crop area estimation can depend on the complex seasonability, complex physiography, cropping patterns, duration of the crops, mixed cropping systems and field size. The sown areas may differ and this can be due to damage incurred as a result of environmental disturbances. In some cases, the biotic factors can damage the crops and hence there will be a reduction in the sown areas. We can consider both biotic and abiotic factors in this case. The estimation of crop area can be possible via multiple phases.

The estimation of crop areas using conventional techniques can be a tedious and costlier process. Satellite remote sensing can provide us images that can be useful in the classification of crops and their yields. LACIE was started by the united states to measure the terrestrial plant ecosystems globally using satellites (Erickson, 1993). In the Landsat program, some improvements were made. The area sampling frame was improved and regression estimation was made (Hanuschak et al., 2004). One of the important factors that aid remote sensing in the identification of crops and their appropriate classification was that the crops appear differently if we consider multispectral images. This can be attributed to the varying phenology, cultural practice, physiology and structure. Ground truth data can be considered as an important factor as the accuracy of the crop classification is based on this data. The estimation of crop area can be done via four approaches. They are calibration estimator, regression estimator, direct estimation and ASF design. The proportion of the crop can be estimated via digital classification and visual interpretation. This can be a basic nature of spatial variability. The area frames can be classified into two types. They are area frames with a regular shape or grid sampling and area frame with boundaries that can be natural or physical in nature. We can use the coefficient of variation in the stratification process. The analysis of images can be done using sample segments and, in some cases, we need to take the entire scene. Using complete scene and estimation can be called as complete enumeration. Whilst considering the sample segments, the area of the segment will be aggregated post the statistical treatment. With this technique, we can get the total area. In the case of complete enumeration, we can use an administrative boundary and images. Then, we can count the pixels that represent crops. The value obtained will be multiplied with the size of the pixel to get the area of the crop. The classification process can be supervised or unsupervised. In the case of supervised calculation, we can use the ground truth data to make classes or classify. In the case of unsupervised classification, we can use the clustering process. In the case of supervised classification, we can use algorithms like Maximum Likelihood or ML, Parallelepiped and Minimum-Distance-to-Means algorithms. In the case of unsupervised classification, we can use K means and ISODATA.

Hierarchical classifiers or decision tree classifiers, Fuzzy-set classifiers, Artificial Neural Networks and Support Vector Machines can be classified as the latest classifiers with new improvements. The classification process will be done using SPOT VGT, AWiFS data or MODIS data and this data can be classified as multitemporal moderate resolution data. We can get single date data with high resolution using Sentinel 2 MSI, LISS III and Landsat OLI satellites. MODIS sensor can work in Terra and Aqua modes. It is onboard EO1. It can be classified as freely available data. We can get 1km, 500m and 250m products. SPOT 5 satellite has two sensors called as HRS and VEGETATION 2. We can get 20m, 10m, 5m and 1 km products with these sensors. VEGETATION 2 sensors provide 1km products that can be freely available. Landsat 8 satellite has an OLI sensor that can provide us data that is freely available. The Resoursesat-2 satellite has three sensors namely AWiFS, LISS III and LISS IV. We can get 56m, 24m and 5.8m products. Some of the LISS III and AWiFS data can be freely available. Sentinel 2 satellite has an MSI sensor and the data from this sensor is freely available. Rapid Eye has a multispectral sensor and it can provide us basic and ortho images. RISAT 1 satellite has C-Band SAR and we can get MRS products with 18m resolution. Sentinel 1 has C-Band SAR and the data is freely available. Radarsat 2 has a C-band SAR sensor and it can provide 30m products. COSMO-SkyMed has X-Band SAR.

Multidate data can be considered as an important asset in the crop monitoring process. It can be associated with the crop phenology or growing patterns of diverse crops that can inhabit the same area. For this purpose, we can use moderate spatial resolution but with higher temporal resolution. This can be obtained by MODIS. It can give 250m products daily and even we can get 8-day products. We can also obtain data from SPOT VGT. It can provide 1 km product daily or we can also get 10-day products. A similar facility is available with AWiFS data. We can get 56m products and for every 5 days, it can provide us data. We can use either 7 dates or 10 dates data that is coregistered in the growing season to classify the crops. In this case, we can use hierarchical classification to classify the crops according to their phenology.

Single date data analysis can be used to know the maximum growth of vegetation of the target crops. This must be supported by the ground data. The target crop can be demarcated from other crops and LULC classes or Land use-land cover classes. In this case, we can use high-resolution satellite data. The satellites that can provide us high-resolution data are Sentinel 2, Landsat 8 and LISS III data. In the development of spectral signatures or crop signatures, we need to use ground truth data. In the case of supervised classification, we can classify pixels of a particular crop using different classifiers. These pixels can be specific to a particular crop. Mostly, the area of a pixel will be multiplied by the number of pixels and this can give us the area.

In most of the South Asian countries, paddy rice can be grown. In the optical remote sensing process, it is tedious to get the images whilst cloud cover. In that case, we can use microwave SAR data to get the image. Most of the SARs can be sensitive to the roughness of the surface. The transplanted rice can give us a low backscatter rate. This can be due to the specular reflection of the standing water. Once the plants grow, there will be increased backscattering. In the case of rice crops, we can expect volume scattering. This can be due to multiple reflections between the water surface and plants. After this stage, the backscattering rate will remain constant. It is important to know that the estimation of rice require data from three different dates. These dates can be prior to planting, whilst planting and post-planting. We need to keep the interval as 20 to 25 days between each acquisition.

The processing of SAR data in the case of rice crops will follow certain steps. They are image georeferencing, image calibration, coregistration, pixel digital numbers to be converted into backscatter values, overlaying and then identification of sites where rice is being grown. We need to devise a decision rule. This will be based on the temporal profile of the backscatter values. Then we can go for the classification of rice crops and their estimation. Ground truth can be considered as an important factor in the classification of crops. We can use this data as an input in both classification and validation processes. The ground truth data can be important to know the land use and land cover dynamics.

The accuracy estimation of the data procured and interpreted can play a major role in the success of the project. For this, we need to know about the confusion matrix. The confusion matrix can be called as an error matrix. This can be used to visualize the classification performance. The confusion matrix can be used to know the internal accuracy of the classification done. This can be determined with the aid of correlation coefficient, Root Mean Square Error and relative deviation. The calibration estimators can be added with the secondary information in the estimation process.

3.3 ESTIMATION OF CROP YIELD

The crop yield forecasts can be important in maintaining agricultural practices. It is important to maintain accurate analysis always. An early warning system can be called as EWS. In this system, it is important to monitor data using collection and analysis methods. The communication of information between the end-users can be important. The association of climate with the vegetation cycles that are recurring can be called vegetation phenology. It is considered as an important variable in both atmospheric and earth science fields. It is important to produce accurate models that can explain phenology using tools that can help us predict the exact response of vegetation towards climatic conditions. NDVI or Normalized Differenced

Vegetation Index can be used to estimate the phenology of land cover types. The slower changes can bring about serious problems in the environment. It is important to go for preventive measures as soon as possible. These environmental changes may need certain warning systems that can forecast climatic problems. A forecast can be considered as an important aspect of any warning system. This technique can be useful in maintaining food security. Important areas of forecasting are biological utilization, access, stability and availability.

These forecasts can provide decision-makers sufficient time to tackle the anticipated problems and plan accordingly. A crop simulation model can be considered as an important tool in forecasting yields. These models can provide us certain information about the weather patterns, soil conditions and management practices. These models can be classified into functional models, mechanistic models and statistical models. Most of these crop models can provide us the variables that are affected by the real-world constraints. Some of the factors like nutrients, tillage, insects, disease and weeds may not obvious whilst using these models. They play an important role in agricultural outputs. These models can help us in framing the decision support models. Most of these models are simple. These models can be used to estimate the yields. With these techniques, we can generate statistical information about the climate and other factors. It is important to analyze climatic data along with meteorological data so that there will be real-time information extraction of the crops. We can get both quantity and quality parameters. The crop simulation models can be based on agricultural statistics, remotely sensed data, water holding capacity of the soils, agrometeorological data and meteorological data. We can develop several indices that can help us in framing the decision support systems.

The results can be linked with the crop yields using regression techniques. This can be called 'model calibration'. We can also call it a 'yield function'. It can be used to measure the yields based on the model outputs. We can use crop statistics in order to calibrate regression models. The agricultural statistics can provide is crop yields and it will be based on trends, weather factors and indirect effects. The indirect effects can be due to weed competition, diseases, pests and weather. It is important to know that soil fertility plays a crucial role in crop spatial variability. Most of the levels in the food chain can be altered due to weather modification. The socio-economic situation can be classified as an important factor that can affect crop production rates. In some of the developing countries, there will be no trend as such. The weather patterns can threaten agriculture in several areas and affect millions of people residing in these areas. Remote sensing data can be used to get information about the physical objects and we can analyze that information for forecasting and other purposes. We can use NDVI, Green Normalized Difference Vegetation Index or GNDVI, Photochemical Reflectance Index or PRI, Normalized Difference Red Edge or NDRE, Canopy Chlorophyll Content Index or CCCI, Ratio vegetation Index or RVI, Enhanced Vegetation Index or EVI, etc.

Chapter 4

Climate Variability and Crops

CONTENTS

4.1 INTRODUCTION

There are two important reasons why the climate differs with the place. The energy that will be received from the sun can be considered as the first factor that can influence the climate. The second factor is the circulation of oceans and the atmosphere. The heat and moisture will be transported with the changes in the atmosphere and migration of the atmospheric entities. The angle of the sun can be considered as one of several factors that can influence both regional climate and global climate. In the equatorial regions, the angle of the sun will be straight. The sun angle at the equator can produce excess heat at these regions and hence the climate variations. The regions that are at higher latitudes can be considered as having a cooler climate than the lower latitudes. There is some vegetation that can get spread all over the land. This is only possible for some plant species. The biomes can be affected by the local climatic patterns. Most plant species can be affected by climate change patterns. The climate system encircles the planet and so there will be swirling fluxes of mass and energy. These fluxes can be via land, water and air. They can be regular or partly regular. The seasonal fluctuations and diurnal changes can be important in understanding the effect of climate on agricultural products and others. The ocean-atmospheric phenomena can be considered as an important factor in determining the climate of the tropics and weather events. The largest interaction between atmosphere and oceans is called El Nino-Southern Oscillation or ENSO. There will be a shift in the pressure gradients. There will be the North Atlantic Oscillation and Arctic Oscillation.

The climate system can comprise of both energy transformations and fluxes. The energy transformations can be studied in the form of momentum, latent heat, sensible heat and radiation. The changes and transport of matter can affect the atmosphere, landmasses and oceans. All the climate system can act as a machine and this will be run by infusion and redistribution of energy. The solar radiations that reach the earth's surface can influence the climate systems in several ways possible. The energy can get transported to all the abiotic and biotic components of the earth. The oceans and atmosphere can be classified as fluids and any sort of internal movements can bring about turbulence in the normal atmospheric circulation. The aerosols and gases can absorb both long-wave and short-wave radiations. The aerosols can be important in the production of precipitation and clouds. There will be an uneven heating of the oceans and this will bring about several weather events on the terrain. This also will change the heat propagation all over the surface of the earth. The land is about 30% of the earth's surface and the atmospheric and oceanic processes can regulate the weather events that occur on the land masses. The shift in the mean climatic conditions in any region can be called as climate change. The effects of climate change can be prominent in certain areas and the effect may be less. The temporal fluctuations that occur in a region can be called climate variability. There is a considerable effect on the atmosphere due to anthropogenic factors. This can be linked with climate change. The warming trend that can be observed due to anthropogenic emissions can bring about climate change.

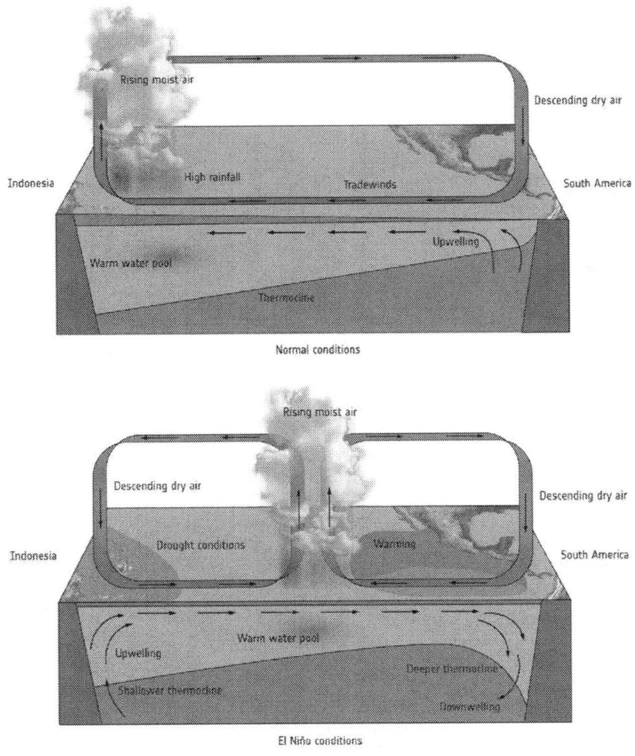

Figure 4.1: El Nino.

Source: https://www.esa.int/ESA_Multimedia/Images/2018/08/El_Nino

El Nino (Figure 4.1) is a Spanish word that means 'the Christ child'. This name was given by the fishermen of Peru and Ecuador. During the occurrence of El Nino, there will be lesser fish numbers and so most of the fishermen will be at loss. This condition can be due to warm ocean currents that can displace the normal cold water currents that are dominating the coastal regions. In some cases, the warm waters can remain for more than four months or more resulting in heavy rains.

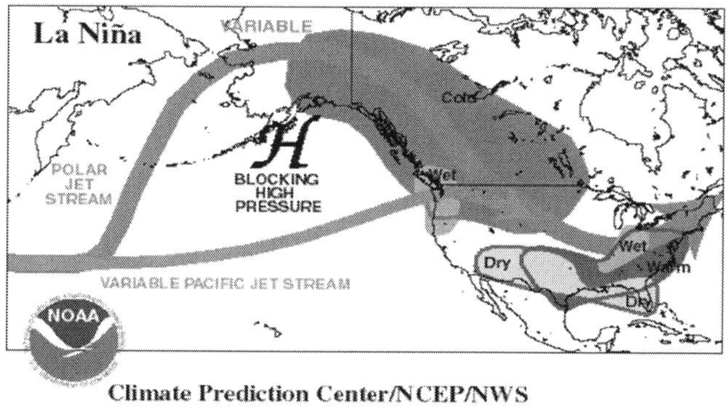

Figure 4.2: La Nina.

Source: https://www.weather.gov/iwx/la_nina

La Nina (Figure 4.2) means 'the girl child'. This represents the opposite situation of El Nino. The sea surface temperatures can be cold and this will occur in the eastern pacific regions. The wet and warm conditions can occur in the western pacific regions. Thermocline can be defined as a layer that can divide the warm water and cold water.

The climate modifications can be linked with ENSO that impacts the agricultural crops being cultivated all over the world. The ENSO events can play an important role in crop production on both spatial and temporal scales. These events can affect the crop production systems that can vary from one place to another. The climate variables, temperature and precipitation can be considered as important parameters that are to be considered to know the crop growth kinetics. The yield variability in the process of agricultural production can affect 3 levels. They are:

- Individual farm
- Farming regions
- Global food systems

These contributing factors can differ in type and scale. The innovations in the crop farming sector can be attributed to the green revolution. This has also contributed to crop variability. The fertilizer enriched crops were sown in order to enhance the crop yields. The crop variations can be attributed to the price modulations, irrigation, agronomic practices, pest infestations, varietal responses, input supplies and ever-changing weather fluctuations. Advances in crop genetics and adaptation of the latest farming techniques can be considered as important aspects in the generation of improved crop yields. The genetic enhancement of the crop varieties can enhance the production uniformity and so there will be increased adaptability to the signals derived from seasonal variability. This will permit crops in order to survive across diverse climates, latitudes and rangeland sites. It will also enhance crop tolerance towards rough soil conditions. Most of the breeding activities can be focused on enhancing the resistance of crops against heat, cold and drought along with climate extremes. Most of the traits that are linked with the hardiness can be attributed to the crop breeding activities. Most of the breeding methods can be focused to reduce the crop yield variability. The improvements in the agricultural practice must be introduced to improve the crop situation.

The crop yield variability can be reduced and this can be attributed to the wider application of pesticides and fertilizers. If soil fertility is high, most of the plants can enhance resistance. Most of the modern agricultural practices like modern irrigation can bring about yield stability. The interaction between the variability factors can be complex. Agricultural intensification can influence variability and productivity. Most of the meteorological events in the growing season can induce detrimental effects on crop production rates. The seed germination stage can be considered as the most vulnerable state in which the crop losses are high. Most of the yields will get reduced and this can be due to crop lodging, pest infestations, water logging and nutrient leaching. Most of the crops will get collapsed due to hail, rain and wind. This can be called crop lodging.

Precipitation can be classified as an important source in maintaining the moisture levels of the soils. The soil moisture levels can determine crop productivity. ENSO affects the seasonal variability of the precipitation. The precipitation events can differ in persistence, intensity and frequency. The rainfall distribution according to the seasons and growing periods can be considered as important factors. The intensity along with the duration of the stress can be important for crop growth. The water stress can be associated with reduced cell turgor pressure, dehydration of the tissues, cell volume shrinkage, and enhanced solute concentrations (Larcher et al., 1965). Water stress can inhibit photosynthesis, chlorophyll synthesis, protein synthesis, cell wall formation and cell division. In the case of extreme stress conditions, there will be increased respiration rates, disrupted plant metabolism, accumulation of pralines and sugars (Ehlers et al., 2003). If water stress develops, most of the plants exhibit reduced osmotic potential that is crucial in maintaining turgor pressure. This osmotic adjustment can permit the growth and enlargement of the cells. Drought can reduce the expansion of the leaves and so their longevity will be compromised. The photosynthetic rates will come down and hence plants cannot live. If drought situations prevail in the reproduction stages of the plant, then the effects will be detrimental. ENSO can affect the precipitation timings and hence most of the crops will get damaged. In maize crops, the pollination stage can be considered as a crucial stage and if water stress prevails in that condition, the plants may lose their yield. Water stress can bring about problems in heading stages and this can be called as 'milk-ripening'. If this happens, there will be damage to the grains.

Drought conditions can bring about senescence in aged leaves and so we can expect shedding of leaves prematurely. There will be moisture stress in the grain-filling, pollination and flowering stages in some cases and this will be detrimental. This situation can be observed in sorghum, wheat, soybean and maize crops. We can categorize drought as rain shortage. The drought condition can be classified into four types (Wilhite et al., 2017). They are

- Meteorological drought (Figure 4.3)
- Hydrological drought (Figure 4.4)
- Agricultural drought (Figure 4.5)
- Socio-economic drought (Figure 4.6)

Enhanced precipitation can result in water logging and leaching conditions of soils. This can lead to enhanced pest infestations, crop lodging and depressed aeration. If the soils are flooded, there will be an anaerobic environment and so the oxygen gas may not be available to the roots for the respiration process. Due to floods, there will be flooding injury. It can be identified by hypertrophy, leaf epinasty and turgidity. There will be the formation of new roots. Wilting may occur due to increased flood situations. The activity of the roots can be diminished as the aeration levels will be lowered. If there is excess soil moisture, most of the operations that are needed for farming will be stopped and this will create many problems. There will be enhanced soil erosion, puddling and soil compaction.

ENSO cycles can bring about high temperatures and this will affect the agricultural conditions globally. In some areas, the temperatures will fall below the normal temperature range. Mostly, there will be high temperatures due to El Nino cycles. This situation may get reversed during La Nina cycles. Frost can prevail due to these cycles and hence the crop growth parameters can be hindered. The variations in temperature can affect plants in diverse ways possible. The abnormal soil and air temperatures can affect the shoots and roots that encounter diverse environmental conditions.

Figure 4.3: Meteorological drought.

Source: https://bifrostonline.org/meteorological-drought-assessment-in-north-east-highlands-of-ethiopia/

The temperature fluctuations will be high at the roots and canopies. This will differ from day to day also. Crop metabolism can be affected by the local temperatures. These temperatures can disturb root and shoot biomass ratios, cleavage of photosynthetic products and morphological developments. It is important to know the critical temperature parameters that can get correlated with the latent heat load and average daily temperatures. Different physiological systems existing in the plants can respond to diverse temperature dependencies. The crops can grow well up to the threshold temperature limits and then they tend to lower their productivity. Most of the crops respond well to the high temperatures and this can be evident from the metabolic activities that are visible to us.

Figure 4.4: Hydrological drought.

Source: https://cpo.noaa.gov/Meet-the-Divisions/Earth-System-Science-and-Modeling/MAPP/MAPP-Task-Forces/Drought-Task-Force-III/Drought-Task-Force-II/Research-Objectives-II

Figure 4.5: Agricultural drought.

Source: https://business-review.eu/featured/agricultural-drought-losses-call-for-emergency-state-say-farmers-representatives-29800

Figure 4.6: Socio-economic drought.

Source:https://drmkc.jrc.ec.europa.eu/Portals/0/Partnerships/Seminars/WorksopDRMKC_2017_June/Presentation_Naumann_DRM_Workshop_Varese_2017.pdf

If there is a sudden spike in the temperatures, there will be denaturation of the enzymes and hence the crops tend to react negatively. Enhanced temperatures can affect the respiration and photosynthesis process negatively. The wheat crops can tolerate temperatures up to 30°C. If the wheat crops were exposed to these temperatures for 8 hours, we can expect reverse vernalization. Rice crops can grow well up to 35°C. At one hour of exposure, there will be spikelet sterility. The maize crop can grow luxuriantly up to 36°C. Prolonged exposure to these temperatures can result in loss of pollen viability. The soybeans can recover from stress and this can be considered as a critical period whilst developmental phases. Potato crop can grow up to 20°C. At these temperatures, there will be tuber bulking and tuber initiation. Cotton crops can grow up to 40°C. If these crops are exposed to 6 hours, there will be an abortion of bolls.

Damage to crops due to extreme temperatures can be based on the developmental stages at which they are exposed. Mostly the reproductive stage and vegetative stages can be classified as vulnerable stages of damage. The extremely hot temperatures can damage the reproductive stages and so they are detrimental for crops. The sensitive stage in the wheat crop is the grain filling stage. In the case of the soybean crop, the flowering stage can be considered as sensitive. In the maize crop, the tasselling stage can be classified as vulnerable stage. The high temperatures can also induce sterility. The high-temperature regimes can cause plant injuries. The injuries can be caused due to excess moisture stress also. The function and growth of the root systems will be hindered due to heat-related stresses. The uptake of water and nutrients will be disrupted due to severe temperature variations. The translocation of the carbohydrates will be halted due to high temperatures. The sink strength of the plants will be reduced and so the fixation efficiency of carbon molecules will be reduced. There will be increased respiration rates. Most of the metabolic reserves will be exhausted due to exceeding respiration levels. It is to be understood that temperature can be a major contributing factor in agricultural production. The low and high temperatures can enhance phenology resulting in faster maturation rates. In hot seasons, the growth cycle will be reduced and so the yield potential. The yields and accumulation of dry matter can be classified as important determinants of agriculture. If there are low temperatures, the reproductive and vegetative developmental stages will be prolonged. This will definitely increase the harvest per plant. There will be an enhanced dry weight. Increased temperatures can yield low productivity rates.

The plants can release excess heat in the form of long-range radiation. The latent heat can be released in the transpiration process (Gates, 1990). In the case of drought, the stomata will close and this will help the plants to reserve water proportions. The stomatal closure will lower the transpiration rates. These adjustments will increase the temperature of the plants. It is obvious that heat stress and water stress can combine to bring about detrimental effects on the crops. Most of the droughts will get surfaced due to enhanced solar irradiance. This condition will be associated with dry winds and warm weather conditions. These conditions along with other factors can result in crop failures. This can be attributed to the decrease in rainfall.

4.2 AGROECOSYSTEMS

Any agroecosystem (Figure 4.7) can be classified as a complex community that has several organisms living in that. It comprises the following three levels that are otherwise called as trophic levels.

- Crop level
- Pest level
- Pest-Predator level

These trophic levels can interact with each other. The magnitude of interactions can determine the rate of infestation if any. It is important to know that humans are one of the important entities in an agroecosystem.

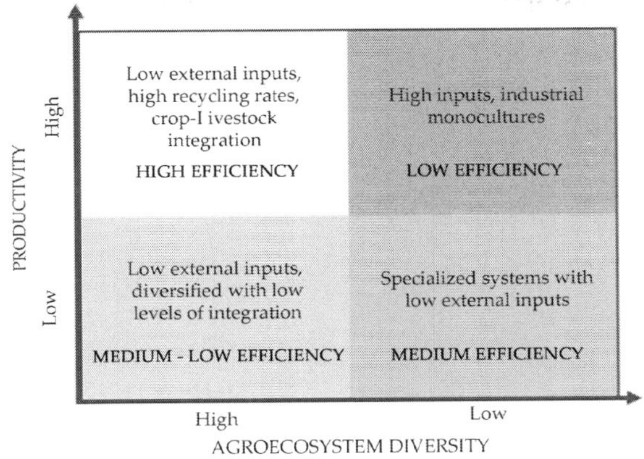

Figure 4.7: Agroecosystems.

Source: https://www.researchgate.net/figure/Features-of-green-agroecosystems-of-the-future-productivity-diversity-integration_fig1_233435738

The anthropogenic activities can affect crop productivity negatively in some cases (Giampietro, 2013). Humans can introduce insect pests into new areas and help in the spread of the pests. The spatio-temporal extent of the pests can get extended. The adaptability nature of the agroecosystems is high and so they are still existing against all odds. They can adapt to different levels of climate variability. It is tedious to measure the factors behind the reduction of the agroecosystems due to climate variability. Mostly, all the agroecosystems have many species along with the crops. The soil mycorrhizae and pollinating bees can be considered as beneficial species for any agroecosystem. The nematodes, weeds and other pests can feed on the crops and hence can jeopardize the crop productivity rates (Peterson et al., 2010). The climate variabilities not only affect the crops but also influence the pests that feed on these crops. The pests can be microbes and organisms that can damage crops and lower the value of crops post-harvest. The major pests we observe in the crop fields are pathogens, insects and weeds. The distribution of insects, pathogens and weeds can be affected by climate change. There are several crops that got affected due to climate variability (Rosenzweig et al., 2001). Climate variability can affect the application timings and can decrease or increase the effect of herbicides. The difference in metabolic activity, chemical toxicities, humidity, soil moisture, wind, precipitation and temperature can affect the crops and weeds. The increased temperatures can increase the growth rates in plant species. This will enhance the speed of the chemical absorption of plants. The rainfall intensity can affect the transport, persistence and effectiveness of the pesticide action. This is obvious that climate variability can bring about several changes in plant growth and development. In the history of agriculture, there were always outbreaks of diseases and pests.

Many researchers have worked to know the association or relationship between climatic variability and pest infestations. In the dry years, the grasshoppers can damage the crops effectively than in other seasons (Smith, 1964). During drought seasons, the viral diseases were high in the crops and the aphids were classified as vectors to carry these viruses (Bagnall, 1977). Most of the pest infestations can be linked with the changes in air temperature and precipitation rates. Increased moisture and humidity can enhance fungal infections and so the crop yields can be drastically reduced. *Phytophthora infestans* can attack leaves if there are high moisture content and humidity. Rice leaf blight, wheat stripe rust, wheat scab and gray leaf spot of maize were caused due to these conditions. The transport of soil-water can also enhance the dissemination of pathogens of soil. Soybean sudden death syndrome can occur due to the spread of pathogens via soil-water transport. The indiscrete soil saturation can cause common smut and cray top in maize crop. The water stress conditions can reduce plant vigor and disrupts carbon and nitrogen ratios. The plant resistance against fungal pathogens, insects and nematodes can get lowered. Water stress can help in the propagation of nematodes and hence lead to an outbreak. The dry conditions can enhance the growth rates of insect vectors and so there will be viral epidemics. Summer locusts and aflatoxin of Aspergillus can affect the crops in dry conditions. Warn winters can also promote the growth of insect vectors and pests. The European corn borer, wheat rust and wheat scab can attack crops in the warm winters. Aphids can damage most of the crops in warm winters. Air currents play a major role in the propagation of disease agents like fungal spores. Leaf blight of maize can be caused due to the propagation of disease agents along with the air currents. Stem rust fungus can occur due to propagules spread by the air currents.

There are several adaptive techniques developed by the plant species to combat the diseases caused due to climate variabilities. The adaptation techniques can occur at different levels. These techniques can be at the farm level, regional level and also in the agricultural sector. There will be several management options available to reduce the water-scarce conditions at the farm level. The diversification of the crops can bring certain challenges making pest management hard to achieve. Better maintenance and lowered tillage rates can lower the yield variability. We can reduce the effects caused due to drought by using rapid yielding varieties. These varieties can withstand drought situations by using limited water and other reactions. It is important to use plant cultivars. There should be an efficient use of irrigation so that the effect of drought can be minimized or even eliminated (Hillel, 2003). Water needed for the crops can be extracted from groundwater and surface waters. Irrigated agriculture can be economically feasible than other crops. It is important to know ENSO cycles so that the irrigation managers can adjust the problems associated with them. It is essential to use breeding crop varieties that can withstand drought situations (Rao et al., 2005). We need to engineer plant species that are a combination of traits that can induce dehydration tolerance and dehydration avoidance (Muchero et al., 2008). The dehydration rates can be avoided with simple modifications in the morphology of the plants via genetic engineering.

Agriculture can be affected by ENSO and this was proved several times. Most of the countries face problems with the climatic situations provided by ENSO. It is important to know that we cannot completely link the devastating crop situations to ENSO cycles. The technology being used, cultivar types, crop types and fertilizer applications can be considered as additional factors that can contribute to crop damage and propagation of pests. We cannot isolate or eliminate the effects caused due to ENSO as most of its effects are detrimental to crop life. There are several variables that can regulate temperature and precipitation. Most of the ENSO effects can be indirect. Decreased and increased rainfall can be linked with cloudiness (Hansen et al., 2008). Most of the water resources can be affected by the ENSO cycles. The frosts that occurred in Australia are due to ENSO cycles. The ENSO effects can be firm and they can affect the commodity prices. There will be rapid fluctuations in agricultural production (White, 2005). The consumers can be affected by the price deviations and so there will be lowered food availability. The relationship between agriculture and climate change can be explained by ENSO events.

4.3 MODELS

The impact of climate can be associated with diverse contexts of agriculture. The biological, ecological and physical factors can affect the agricultural production rates in both developed and developing countries. The effects can be economic, political and social. It is important to know that the effect of climate on agriculture will be comprehensive in nature. There are several methods and models that were developed to know the relationship between climate variability and agriculture. The conceptual models can be considered as important as they can guide us in the research initiatives. Social scientists, biologists and physicists can use these models to predict climate patterns. The primary step in any modeling process is to define agricultural systems. The regional analysis using these models can be important. In the process of selecting an agricultural system, it is important to identify the users and the extended scope of analysis. This can be done by framing economic analyses, biological analyses, physical analyses, scoping exercises and planning exercises. There should be an interaction between NGOs, agricultural planners, input providers, climatologists, meteorologists, agronomists, framer groups and farmers. It is important to identify and decode the stress signals from the agricultural areas. We can categorize important sources of climate variability. For this, we can use statistical tests. There should be an interaction between agronomists, climatologists and meteorologists at regional levels. We need to analyze social and economic aspects. These studies must be done at national levels, regional levels and household levels. In this process, we can use social science methods. In these methods, we can use surveying methods and interviews at household levels. The economic analyses can be important. It will help us to know the supply and demand needs. We can forecast the risks and other economic concepts in this process. In this step, we need to interact with planners, economists, sociologists, anthropologists and rural households.

It is also important to know both spatial and temporal patterns. In this step, we need to consider sea-surface temperature, decision-making calendar and crop-growing seasons. At this stage, it is important to use remote sensing data along with the ground truth data. We must analyze the regional climate and agricultural monitoring devices. In order to achieve this, it is essential to improve GIS tools and calibration as and when required is important. It must be ensured that crop models should be validated from time to time. We need to interact with the farmers and agronomists. It is important to develop and maintain advanced seasonal climate prediction tools and models. We can focus on downscaled forecasts like weather generators, climate models, regional climate models and GCMs. For this purpose, it is essential to interact with the climatologists and meteorologists.

The risk management strategies should be tested so as to ensure that if there are any adaptive risks laid within. We need to search for opportunities that may get surfaced whilst the modeling process to begin. For this, it is important to resort to crop models associated with GIS simulations. We need to use risk-pooling estimations, participatory methods and ethnographic methods. It is essential to use several simulations so as to keep the error rate at a minimum. There should be an interaction between the local banks and farmers so that constraints if any can be highlighted and addressed there and then. We need to create an adaptive management strategy to tackle climate variabilities. For this, we can use the GIS-based decision analysis. We can use index insurance programs. The research groups and users should be supplied with continuous feedback. This feedback can be used by the institutions.

In order to know the decision-making process, we need to go for conducting interviews so that the exact picture can be obvious for us to frame certain protocols. These interviews can extend our scope of understanding. We can know the sensitivity of the farmers and enterprises against climate vulnerabilities. These interviews can be helpful to know the facts about El Nino and La Nina events. The researchers from social sciences, biological and physical sciences can interact with the decision-makers to enhance the climate forecasts and their accuracy. The weather prediction can be different from the seasonal climate prediction (Zebiak, 2019). In the case of weather prediction, we tend to observe the measurements that are based on actual conditions pertaining to pressure and temperature. These observations and measurements can be based on numerical models. The actual and initial conditions can help us to know the forecast for weeks to come. These predictions can be called as short-term predictions or forecasts. These predictions can be accurate and maybe best useful for few days. These weather predictions are vulnerable to climate systems and so we will be forced to modify the numerical models according to the new normal. The accuracy of these predictions can be based on the tools that are to be used to know the present conditions and the personnel deployed for this purpose. The measurements cannot be accurate as the predictions are subjected to several elements that may or may not influence the actual conditions. The interannual ENSO forecasts can be done using dynamic models and statistical models. It is always important to predict SST or Sea-Surface temperatures in the Pacific regions. This data can be used in framing global climate models or GCMs. There are several approaches that are simple. We can use lagged statistical relationships to predict global rainfall using the Southern Oscillation index (Stone et al., 1995). Forecasting of SSTs can be considered as an important and primary step in this process. We can use data obtained from the tropical pacific regions. We can use dynamic models. We can use SST for three months or six months. In this process, we can use data from the tropical Atlantic Ocean and the Indian ocean. We can use statistical CCA models for this purpose also. The second phase will be to forecast the climate. This can demand data from global and regional scales. We can use atmospheric general circulation models for this purpose. We need to use all these models to achieve an accurate forecast that is valid on different fronts. In some cases, we can use statistical regression models and composite analogs for this purpose.

It is important to define models and rectify biases generated during the acquisition and analysis of data. We need regional data for this purpose. We need to use statistical methods in this case. In this phase, we can compare the forecasts of the past seasons and we can generate consistent models with reduced biases. It is important to create forecasts. The data can be collected from both regional and global scales. We can use objective weighting factors for this purpose. It is important to combine the results and add all the possible scenarios to eliminate biases. NOAA NCER, SSES HiDyn, ENSO-CLIPER models can be classified as statistical models. NOAA NCEP model uses the Markov model, Canonical correlation analysis and Linear inverse model as prediction systems. SSES HiDyn model uses a Bayesian hierarchical dynamic spatial model (Berliner et al., 1996). ENSO-CLIPPER uses a Multiple linear regression model (Landsea et al., 2005). POAMA, ECMWF ECHAM and NASA NSIPP can be classified as dynamic models. POAMA used a coupled ocean-atmosphere model as a prediction system. ECMWF ECHAM uses the same model. NASA NSIPP uses a coupled global ocean-atmosphere-land model (Roxana et al., 2004). LDEO and SCRIPPS/MPI can be classified as hybrid models. LDEO used intermediate coupled ocean-atmosphere model whereas SCRIPPS/MPI use hybrid statistical atmospheric model, dynamic ocean model and atmospheric models.

Forecasting skill can be defined as a tendency to forecast that can be based on a model or via an ensemble of models. In some cases, these models can be used in the form of mixtures. It is important to know ENSO cycles completely to give an accurate forecast. The forecasts can be validated and verified with the experts as and when needed. It is always important to compare the previous forecasts with the present one so that we can validate them effectively. Enhanced skills in interpreting ENSO cycles can help us produce accurate weather maps that are needed for successful agriculture. There are several measurement methods that can be used to evaluate the forecasts. The change in the atmospheric and oceanic conditions can be studied used observed weather and forecast scenarios. They can be based on the conditions that prevail in the geographical extent. It is important to use agricultural tools. Using agricultural tools, we can expect an agricultural response and predicted agricultural response. It is important to opt for adjustment scenarios. This can be used in land use dynamics and technology management. With this, we can even forecast the economic consequences and related damages.

It is important to conduct regional studies pertaining to ENSO and other oscillations in order to predict climate situations. This knowledge is important as climate prediction can duly impact the production rates.

Chapter

5

Prediction of Seasonal Climate and Agriculture

CONTENTS

5.1 INTRODUCTION

Agriculture can be affected by insufficient rainfall or precipitation and these events are linked with climate change and its variabilities. Drought can be considered as an important factor that can damage most of the crops in the shortest time possible. Most of the economies will be dwindled due to irregular weather patterns that were derived from both local atmosphere dynamics and global transfers. The seasonal forecasts can be considered as an essential part of any decision-making communities as most of the world is on the verge of climate sensitivity and it needs to be addressed immediately and effectively. In the primary production stages, weather and climate can be considered as a single management sector. Due to extreme pressure on the resources, producers are shifting towards innovative solutions and products. In Australia for example, the agricultural sector is mixed with both cropping and livestock production along with the interference of climate change. Most of the farms can be classified as capital intensive domains. Some of the crops can be classified as family-operated entities. Mostly, the farmers in most of the nations suffer from cost and pressure dynamics. The fact behind this is that the farmer communities neither can decide nor affect the price dynamics. Most of the farmers were trained to use advanced tools and techniques to address the present and anticipated climate risks. The climate can be variable and it is based on the continent landscape. The scope of farmers is being extended to face the risks that can be due to climate variability. Several scientists focus on modeling and analytic tools to give us an accurate forecast. These models can be put into practical use in several sectors. There are several financial pressures existing in the agricultural sector. Grain growing sectors can be prominent and the determining factor is soil moisture and other associated factors. The soil moisture can be added to the soils due to rains and can be stored in the fallows. This can be considered as important as far as the crop growth rates are concerned. The moisture profile of the soils will drop down if the precipitation events are less. The stored soil moisture can get reduced drastically in the case of drought conditions. In the production of winter cereals, the occurrence of frost can reduce the flowering process and so diminish the yields. It is important to know the accurate forecast so those wise and appropriate decisions can be made. It can be useful in deciding the plant variety that is to be planted. The rainfall events that occur whilst the harvest period can push the experts towards modifying their decisions pertaining to the crops. Sufficient rainfall and moisture can enhance crop yields. We can use more nitrogen fertilizers so as to achieve high protein content. This decision must be made prior to planting. In more wet conditions, there will be a reduction of yields and this can be attributed to the waterlogging conditions and diseases. We can expect fungal diseases when there are waterlogged conditions.

The cotton crops can be important as it is a commercial crop that can provide employment to many. In some countries, the cotton-producing lands are less and may not meet the growing demands. This industry may need sophisticated equipment and advanced management skills so that there will be a smooth adaptation shift. There are several reasons why there is a need for expensive chemicals to bring about higher yields. It is the communication network that can help them to adopt the latest technologies into the agricultural domain. It is the dryland farmers who grow cotton that are the first to realize the impacts provided by the El Nino cycles. They realized the maximum use of moisture and seasonal forecasting from time to time. These techniques and information provided them to use the of irrigation systems with the aid of stored water. This has led to the installation of rain harvest systems that can hold flood water and the rest can be stored in dams.

The pest and weed problems can be affected mostly the climate fluctuations. The insect pests can proliferate in specific weather conditions ad this can be dictated by other factors too. These infestations can bring about outbreaks that may be more than expected aircraft to control. In many cases, the weeds can grow well in wetlands even when crops are compromised. If we have a seasonal forecast with us, we can plan the spray applications in advance. The prawn industry is another important industry that affects due to seasonal variations. Correlation between the El Nino and prawn cultivation has been established. When the stream flows become slower, the prawns may not grow well and this can lead to reduced harvest. Climate variability can have a profound effect on most aquaculture plants.

Weather can change and is never constant as it follows no standard parameters. The climate variabilities can be due to anthropogenic factors and if this continues, we may face a lot of problems. Some places may get dry conditions and this can be attributed to climate change situations. There are many climatic variations that can be classified as coherent on a larger scale. It is important that the workforce must be skillful to extract the information from the weather variables. It is important that the observers should have knowledge of sea surface temperature or SST. The most energetic interannual variability can be observed in the case of ENSO. The ENSO prediction studies can be used in advancing the ocean-atmospheric modeling. This has helped us to build a platform for ocean observation. It is important to know that the tropical pacific regions can be called as the origin for ENSO. The ocean-atmospheric interactions in that region can be classified as the generating tools for climate variabilities. The tropical Pacific regions can be classified as source regions mostly. The ENSO prediction systems are based on two problems or parts. The ENSO prediction as a function of global influence is considered as an important part. It is essential to predict the impact of ENSO on economics, health and agriculture. Appropriate strategies should be formulated accordingly. The International research institute for Climate prediction can provide us enough information about the climatic impacts on agriculture also. It also gives us information about the climatic variations. These climatic variations can bring about unprecedented changes in defining social issues. For every four years, there will be a warning event that can be associated with the ENSO cycles. This warm event can be considered as irregular as it is being disturbed by anthropogenic events. We can observe changes in its amplitude. In the western Pacific at equatorial regions, we can find abnormal heating of the surface waters.

Due to these warmer surface waters, we can expect enhanced convective activity. There will be a disruption of the trade winds. These changes can bring rainfall to the Ecuador and Peru regions. Generally, these regions were classified as arid regions. Severe droughts can be observed in southern parts of Africa and Australia. We can even expect cyclones over Hawaii and French Polynesia. The Indian monsoons will get disrupted due to these changes. The regional climates of Africa, North America and East Asia will be affected. The seasonal rains in the northeastern parts of Brazil will be disrupted. ENSO event occurred in 1982 and it was considered the worst environmental disaster in history. It left the equatorial waters of the South American hot by two degrees. In the coastal regions, the temperature was increased to 6 degrees. Most of the trade winds got reversed. Due to these events, there was a lot of damage to the economies residing in the coastal regions. Many researchers classified El Nino as a massive warming event of the coastal waters off of Peru and Ecuador. These warming waters resulted in the killing of guano birds and fish. These events have devastated the local and native economies. Flooding resulted in the uprooting of several households. The atmospheric proportion of the ENSO event was observed late and so these consequences were linked with this phenomenon was studied late.

The natural and agricultural ecosystems can be linked with government systems and business. Some of the water resource systems can get affected due to climate change fluctuations. Most of these systems can be classified as dynamic and they can respond to climatic variations. The policy issues, conservation, profit and production and policy issues can help us focus on these systems. The risks caused due to climate variations can bring about environmental losses. This is important in the decision-making process. The climate forecasting skills can be considered as important. The forecast information can help us to alter the decisions as and when needed. The forecast information can affect the outcomes of any policy. It is important to understand the systems approach in the climate forecasting process. This approach can be generalized. This will help us to know the feedback from the climate forecasting. We can go for evaluation and implementation is a better way. We can evaluate the tactical decisions and options if any. We need to determine the opportunities that can be needed in the case of seasonal forecasts. We can understand the impacts caused due to climate variability. We can even understand the systems and their management. Apart from the role of biophysical ecosystems, we can understand the management of these ecosystems in climate risk environments. The government systems can be responsive and dynamic and this also true with climate sensitivity. Climate variability can produce risks in the process of decision-making and this can be attributed to several factors. The seasonal forecasts developed in the last decade have helped the research communities to study the regional climatic conditions in detail. Due to this, people starting using the risks positively and were successful in generating markets (Hammer et al., 2005).

Any system can be classified as a network in which there will be an interaction between the networks, inputs and outputs. The dynamics in the system can be considered as a prevalent condition. This can be applied to both natural and agricultural ecosystems. Several government and business systems can be linked to these dynamics. Defining boundaries can be stated as an important task when dealing with system environments. The internal component, external component and scale of the system can be classified as important entities in any system. These issues must be considered whilst dealing with system dynamics. The function and management of the system can be based on several factors. There are certain systems that can be associated with the natural and agricultural ecosystems and linked sectors. They are species population, catchment, national drought policy, marketing, farm and management and crop field management. In the crop field management section, the chemical, physical and soil attributes can be taken into consideration. We can also include management practices like harvesting, spraying and planting. The inputs in this case can be pesticides, fertilizers, fuel and daily weather. The outputs include soil loss, runoff and harvested products. It is important to consider balances that deal with N and C concentrations. These concentrations may be recorded during the entry and exit from the system.

The chemical and physical processes that occur in the soil can be classified as dynamic parts. In the system management potion, we can manipulate the system and this can be done to achieve targets that end-users need. The managers should be able to understand all the elements that make a system. The manipulations that are needed in the management process can be used to evaluate the system models. The efficiency of these models can be checked. In this systems approach, the decisions about the management and policy pertaining to the natural and agricultural ecosystems can be dealt with. It is essential to use the systems approach in the decision-making process so that we can check its efficiency. This concept is present in most of the rational approaches that can be classified as problem-solving approaches (Robertshaw et al., 2000). These approaches can be assumed as the foundation for the development of the analysis in the agricultural sector and decision analysis (Dent et al., 2019). An essential approach to solving a problem or problems can be classified into some steps. They are:

- Definition of a problem or issue in the systems,
- Defining or determining the objectives/ criteria,
- Collection of information and research products of the system,
- Identification followed by evaluation of the options pertaining to the decision analysis,
- Implementation of the decisions and monitoring the outcomes.

It is always important to define the system where problems exist. The declaration of the objectives can be important in the problem-solving process. We need to go for credible system models that can help us in the problem-solving process. With these steps, we can identify the alternatives and even opt for simulation analyses. It is always important to know the scientific advancements that were included in the models. The scientific efforts can be important in the development of the system models. It is essential to correlate scientific advancement with the forecasting tendencies in the successful operation of models. The systems approach can be important in solving problems that deal with the connections between the researchers, modelers, advisors and decision-makers.

In the case of producing an application that is effective in proving us seasonal climate forecast, we need to consider forecast models and information pertaining to the climate variables. These applications can be used in decision analysis in predicting problems. It is important to enhance the system performance so that there will be effective modeling and the decision analysis will become easier. The main target of the systems approach is to develop needed information that can affect the decisions in the system we selected. This concept holds good with the issues and scales that are prevalent in natural and agricultural ecosystems.

In the case of seasonal forecasts, we will retrieve the information that can be associated with all the systems. The systems approach can include systems modeling and the extraction of information will be specific. The model starts with the farmer sector and then it is followed by business and other resource managers. These sectors will be followed by the government sector. We have different sub-sections. Some of them are crop pasture hard management, property management, land and water resource management, drought alerts and population management followed by marketing policy. All these are based on the seasonal forecasts. The seasonal forecasts can affect nation, region, catchment, farm and paddocks.

The first thing is that we need to understand the system completely. Knowing the crop type can be important to explain the dynamics. It is important to develop predictive systems and tools necessary. It is also essential to understand the impacts caused due to climate variability. Here, the main issue is climate variability. It is important to know the seasonal forecast as and when needed. We need to define the opportunities that are needed to tackle the management process and this should be as a response to the seasonal forecasts. We need to compare the optical before making tactical decisions. Simulation analyses can be used in constructing the system models so that the technical options can be maintained. The evaluation and implementation process should be participative in nature. This can be achieved by generating insights from the decision-makers. It is essential to go for climate forecasting. Improvements in the climate forecasting process should be considered as an asset. This can help us to give forecasts and can enhance climate research.

5.2 CLIMATE FORECAST

The relevance and value of the climate forecast can be essential for both rural and urban communities (Todd, 2014). Troup's analysis confirms that there can be a seasonal structure. This can be associated with the sea-level pressure fluctuations on at the global scale. The rainfall variability can be considered as an important principle component and it can be linked with the Southern Oscillation. Statistical climate forecasting can be important to understand the nature of pressure fields in the Pacific Ocean and the Indian Ocean. Research in seasonal forecasting can see EOF or Empirical orthogonal functions. With this technique, we can isolate regions that can affect rainfall distribution and patterns. This system can be classified as an essential tool in constructing models. This method can demarcate India Ocean SST and ENSO cycles (Drosdowsky et al., 2006). We can use atmospheric GCMs and complex ocean-atmosphere models (Hunt et al., 2003). There are some models called limited area models that can attract the data from ENSO cycles. Most of the accurate forecasts pertaining to El Nino and La Nina can be provided by these models (Allan et al., 2001). It is important to rely on long time-series data. This can be utilized to predict potential predictors. We need to use statistical methodologies so that we can perform analyses with this data. The analyses power of the computer can be useful. It is essential to know both scientific knowledge and understands the prediction relationships. It is essential to predict the climate using statistics.

The observational data networks can be formulated to give us an accurate forecast about the weather situation. This will be based on the spatial coverage and the wider the observational data stations are placed the lesser the resolution. It is essential to maintain these data stations properly and must be kept operational at all times. Since the number of stations will be installed gradually and as per need, we cannot expect long-time series data for all the stations. We can get the SST data on a global scale from COADS or Comprehensive Ocean Atmospheric Data Set (Slutz et al., 1983. The data obtained from these stations can be interpolated in the case of missing data. In some cases, we can use the values generated from the satellite radiances. This will give us SST values on a global scale (Rayner et al., 2011). There are many sites where we can get large scale multi-variable data sets. This can enhance computing power. With computer-assisted technologies, we can enhance the statistical capabilities. We can use statistical techniques for many purposes. We earlier were using statistical methods like linear correlation and regression methods. In the case of multivariate analysis techniques, we can use data reduction techniques. Some of the important techniques are Canonical Correlation Analysis or CCA, Cluster analysis, Singular value Decomposition, Empirical Orthogonal functions and Principal Component Analysis or PCA. These techniques can be used in climate research and seasonal forecasting (Mann et al., 2013). One of the most important forecasting techniques is LDA or Linear discriminant analysis. We can use neural networks to get seasonal forecasts. The forecast schemes can be statistical or dynamic in nature. We can use all the data obtained for this process and this can be called a model fit. We can even test the models using independent data. The demarcation between these two modeling patterns can be done. We can use a cross-validation procedure.

Much progress has been made in the field of climate predictions and all of them are based on numerical models. There are several modeling approaches in addition to the statistical methods. There are several physical and statistical models that were developed to forecast ENSO. These models can predict prior to the onset of the seasons accurately. ENSO cycles can be linked with closed interactions between the atmosphere and oceans. The climate models help us understand and simulate several aspects pertaining to climate variability that is prevalent in the Southern Hemisphere. It is essential to use computer models that can reproduce the near-accurate behavior of the atmosphere in diverse time scales. The time-scales can be classified as longer trends and decadal. In some cases, we can use synoptic events that represent data for 5 days. The effect of SSTs can be strong at the higher latitudes. The accessibility and performance of the climate models enhanced greatly in the last decade. These models help us to know the climate variabilities on a variety of space scales and time scales. Intercomparison experiments have been made. These experiments were done by many research groups and they found that there are considerable variations in the rainfall patterns in the monsoon periods. These monsoons were affected by SST fluctuations.

5.3 STATISTICAL METHODS

The General Circulation Model or GCM can be considered as an important tool in climate forecasting. The forecast systems generated can be used in the simulation of agriculture and decision analysis related models. In this case, we can use cluster analysis and principal components. This model can be applied to the time-series analysis of SOI or Southern Oscillation Index. We can produce SOI phases or SOI types. The prediction of SOI can be attributed to the GCM simulations on a long time scale. We can use sea-surface temperature or SST data that can be derived from the historical datasets. The simulation models linked with the pastures and crops will need daily inputs of precipitation, radiation, evaporation, etc. These variables can increase the efficiency of the models. We can classify SOI patterns and it will be based on the seasons and years of data derived from various stations. We can call this procedure te lag-SOI phase system (Stone et al., 2005).

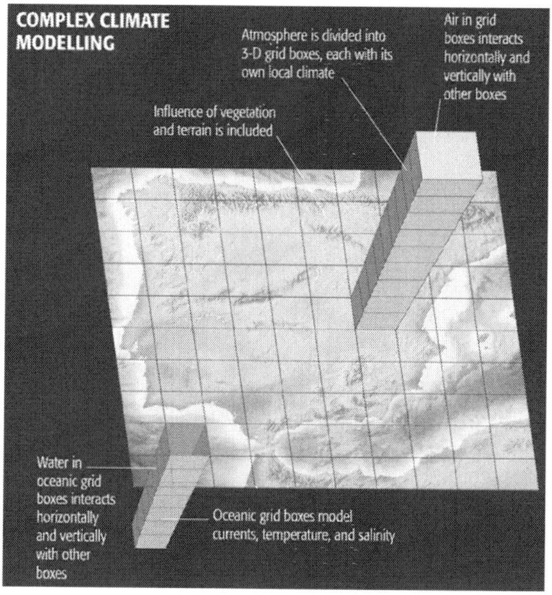

Figure 5.1: General Circulation Model.

Source: https://www.e-education.psu.edu/meteo469/node/140

The SOI can be defined as the difference in the pressure fluctuations in the given regions. There were many improvements in GCMs (Figure 5.1). These models can be used in providing the forecasts related to the SOI, SST and rainfall. The forecast data can use many inputs that are mandatory to build agricultural and crop simulation models. The forecast data can be derived from the GCM models. This can be given in the form of SST fluctuation maps. These maps reflect the precipitation probabilities and spatial variabilities. We can use GISST or Global Ice-Sea Surface Temperature data set. With this dataset, we can derive EOFs and SOIs of the SSTs. In this data, we can get 2^0 grids and earlier, we used to get 1^0 grids only for the entire globe. It is important that we remove the annual cycles prior to EOFs calculation. In general, these annual cycles will spread across 121 years. If we use these annual cycles, then there will be unprecedented domination of this data over other variables.

We can use long-term simulations and even we can consider 5 years of data. In this case, we can achieve a horizontal resolution of 1.9^0. Every simulation can be attached to the GISST data set. The GISST data sets can give us data from Jan 1, 1871. The GCM forecast pertaining to SOI can be used in the simulation. It is important to use principal component analysis whilst using the time series data. We can conduct PCA on a matrix and this may contain forecasts of every year. We can apply the hierarchical clustering technique to the scores obtained from the principal component analysis. We need to identify important break-points. This can be done using cluster dendrograms. With this, we can isolate the GCM of the SOI types or g-phases. These g-phases can be used to detect the analog periods. We can also know the precipitation probability and distributions. After this, we can test the significance of the method being used. These tests can be used to demarcate the rainfall probability and their distribution. The statistical tests can be linked with the distributions of the rainfall probability.

The ability of the g-phases to demarcate these distributions can be based on the cluster analysis and PCA. The forecasting system can be based on the statistics. The EOF scores for every month can be considered for analysis. The cluster analysis can be important in constructing a matrix. These analyses can be important in the pasture and crop simulation models. The rainfall distributions can be generated from the g-phases. These g-phases can be compared with the SST-EOFs. The demarcation in the rainfall probabilities can be important in this analysis. There are many regions in the world that will be affected by rainfall. The rainfall variability can be linked with the ENSO cycles. The farm managers and scientists can detect the decisions that are linked with the forecasting climate. The effective forecast can enhance the farming enterprise. This can be reflected by the increased profits. This can be done with improved sustainability indicators like organic matter, soil and erosion. This can also lower production risks. The frequency of the rainfall can affect crop growth in the drylands. We cannot completely link the rainfall events with the crop yields. The crop simulation models can be used in maintaining the climatic records. These climatic records can be used to establish relationships between the SOI phases and crop performance. The increased rainfall variability can be considered as an important source of yield fluctuations in the drylands. Most of the sub-tropics and semi-arid tropics can be affected due to rainfall fluctuations. There will be lag-relationships between the temporal distribution and ENSO events. The El Nino events can be linked with negative SOI values and this can be extended to one year. It is essential to use an expert or skillful seasonal forecast by the farm managers so that they can make improved crop management decisions. The frequency and timing of the rainfall events can affect the crops that grow in the dryland crops. It is an important task for the farm managers and scientists to notify the decisions that can be affected by climate forecasting. These decisions can be important to enhance the economic performance in the long-term. The farm managers can appreciate the agricultural systems and their interactions in the decision-making process. The seasonal forecasts can be linked with the simulation capabilities. There will be an interaction between the decision-making processes and farm managers. It is the duty of the farm managers to describe the rainfall probabilities and crop forecasting systems to be used.

El Nino's can be linked with the low rate of rainfall events in South Africa, India and Australia. In some parts of South America and other parts of the world, there will be increased rainfall events. Most of the SOI patterns can be classified as 'phase-locked'. This can be locked with the annual cycle. This SOI analysis can give us skill in measuring the rainfall probabilities. In 1997, there was a rapid drop in the SOI values. There was a shift in the Walker circulation and this can be attributed to the increased SSTs. This trend was predominant in the Eastern Pacific.

Whilst managing the cropping systems, most of the farmers will take certain decisions that can be affected by several factors. They always think about the financial returns and it is their ultimate goal. Some of the decisions will be based on lifestyle, soil degradation risk, disease control, cash flow and risk of economic loss. The SOI index can be promoted on the basis of seasonal climates. The SOI outputs have helped the farm managers in taking tactical decisions. The cropping systems can be determined as a sequence of crops that are to be grown on a rotational basis by the agricultural communities. It is important to know that the ultimate goal of any farmer is to get financial returns as soon as possible. This goal can be achieved by taking into consideration lifestyle influences, marketing, political, labor capital, natural resources and economic factors.

Chapter

6

Climate System Models

CONTENTS

6.1 INTRODUCTION

It is important to use high-resolution climate system models as tools in determining the climate variabilities. These models can be used extensively in climate research studies. Most of the economies will rely on sustainable resources. Most of the decisions and policy-making scenarios will be based on the high-resolution climate models. Global warming has become a global problem these days. The projection capabilities of these models should be upgraded so that the farmers will get benefitted. The projection capability at the regional scale is not reliable with these models. It has been an important achievement to develop a high-resolution climate system model that can simulate the climate of East Asia. The monsoons in this region can be due to specialized climate phenomena. The low resolution and middle-level resolution models come with high amounts of bias whilst simulating the seasonal variables. Due to these biases, there will be uncertainties in the models that make them unreliable. It is essential to develop high-resolution climate models to get accurate simulations. Making a high-resolution climatic model has become a challenging task. It is important to enhance the performance of the climate system models so that the climate simulations can be enhanced. The projection abilities of the model should be improved at the regional scales. These models should be sensitive enough to adapt the climate change. We need to install a climate system model that can integrate atmosphere-ocean-land-ice modules.

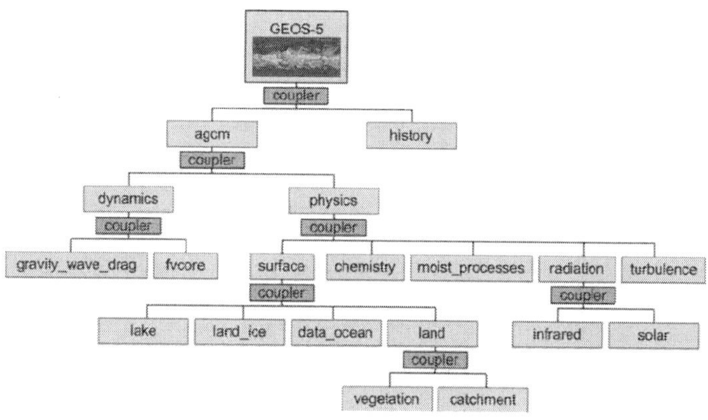

Figure 6.1: Atmospheric general circulation models.

Source: https://www.researchgate.net/publication/220663076_Earth_system_curator_Metadata_infrastructure_for_climate_modeling/figures?lo=1&utm_source=google&utm_medium=organic

Oceanic general circulation models or OGCMs and Atmospheric general circulation models (Figure 6.1) or AGCMs can be considered as an important component of the earth system model that can be linked with climate variabilities. The AGCMs can be classified into two categories. First is the dynamical core and second is the parametrization of the physical processes. In the first portion, we can use Navier-Stokes equations that are related to the atmosphere. We can use several numerical algorithms. The second part can determine the statistical effect of the sub grid-scale processes. The stability and the consumption of time can be important to get an AGCM at high resolution. It is important to build a stable and accurate dynamic core that can bring about AGCMs at high resolution. In any climate systems, the radiative processes can play a major role in explaining climate change scenarios. The parameterizations of radiation can be classified as an important process in framing AGCMs. There are several uncertainties reported in calculating the radiative transfer equations and calculations. There are many dynamic cores that can be used to solve the equations related to the atmosphere that can be associated with AGCMs. It is important to know the dynamical cores so that the model simulations can be done easily (Williamson, 2018).

Climate can be explained in terms of weather states. The probability distribution can be considered as an important aspect of the climate modeling process. There is a fundamental variation between climate forecasting and weather forecasting. It is important to identify the current state of the model. This is essential to know all the details of the model to perform a projection. There can be a simple model that can be called as persistence. We need to know that the weather conditions existing today may not exist tomorrow. It is important to use advanced models instead of relying on numerical models. These numerical models can be called as numerical weather prediction models or NWP models. These models can be used to forecast the earth systems as and when needed. The climate forecasting models use almost similar models. The main aim of climate forecasting is to identify and classify the distributions. These models can be used for a long time so that we can identify the weather states accurately. In the case of using a weather model, there will be two states. In the case of climate, we can get the same distribution and this will be devoid of the model being used.

Any model can be defined as a representation of systems. These models can be classified as building blocks. We can consider a model as an abstract. Models can exist in our natural environment. Some of the abstract models can be mathematical in nature. In physics, we can describe the events that occur in the physical world through a model. Most of the models are familiar to us and they are classified as physical in nature. Some of these models can be constructed with the aid of numbers. Most of the climate models can be classified as numerical models. These models can calculate mass momentum and energy. It is to know that energy will be conserved and so we can measure it. We can consider the climate and weather as dynamic systems. These variables will fluctuate over time. The climate models can be linked with weather forecast models. These models can simulate the movement and interaction of air and water. They can simulate the heat exchange. These models can use fluid flow principles. Most of the models can be used to explain physical objects and their interaction with the external environment. Uncertainty can be considered as an important aspect of the forecasting process.

A model can be considered as erroneous and this can be due to structural errors. This can be called as model uncertainty. Scenario uncertainty can be important in climate change modeling. Some of the variables may not be predicted and so we call them 'parameter's there will be logical dominance of scenario uncertainty. If we predict the uncertainty in the initial phases of the system models, we can call it 'initial condition uncertainty'. It is essential to address all the uncertainties possible in the climate prediction process. We need to understand all three types of uncertainty that can determine the quality of the forecasts. Once we know about all the uncertainties, it will be easier for us to quantify total uncertainty. These uncertainties can be operated in diverse time frames. The first uncertainty called Initial condition uncertainty can exist for a short term. It can occur in weather scales or annual/seasonal scales. The second type of uncertainty called scenario uncertainty can exist for along term. We can say that it exists for centuries if not for decades. The third type of uncertainty can be called model uncertainty and it can operate at diverse timescales. This type of uncertainty can be hidden or can be masked by initial condition uncertainty and scenario uncertainty. Due to the complex nature of these uncertainties, it can be hard in some cases and even sometimes can be easier. Because of this versatile nature, we can easily define uncertainty. It is important to construct a framework so that we can understand the climate predictions in a better way.

6.2 CLIMATE SYSTEM

Climate can be experienced at the earth's surface. This can be classified as an intersecting point for several parts of the climate systems. The components that make up the climate system cannot be completely understood and hence making the simulation process complicated. We need to set certain building blocks that can be assumed as components of a climate system. This can be done by considering the individual processes. For example, the formation of clouds is an individual process that can be linked with the atmosphere component and another set of components called land and ocean. In a climate model, we need to make an attempt to represent diverse interactions that may occur between the components and processes. We can consider three regions of our planet whilst constructing a climate model. They can be the atmosphere, oceans and land. Cryosphere can be added to these regions. The atmosphere can be considered as a primary component in any climate system. We have inert nitrogen, oxygen and other trace gases like carbon dioxide, water vapor and argon present in the atmosphere. Most of the oxygen can be supplied to the environment as a by-product of plant respiration and other species. The oxygen gas can be produced by the plants as part of the photosynthesis process. The oxygen generated will be absorbed or even reacted with ore and rocks of the earth and may be removed from the atmosphere. It is important that these organisms ceaselessly produce oxygen so that it will be available for all the species on earth. We need to consider the trace gases whilst planning a climate model. Methane, carbon dioxide and water vapor can be considered as greenhouse gases. Greenhouse gas can be transparent to the visible radiation that is being emitted by the sun. This gas can be opaque to the infrared radiation that can be emitted by the surface of the earth whilst re-radiation.

If these gases are high in the atmosphere, there would be global warming. Water vapor can be considered as an important greenhouse gas that can affect the climate systems. Carbon dioxide can also produce the greenhouse effect. Solar energy can be absorbed by the earth, clouds and atmosphere and they can reflect some portion of radiation. The ratio of the reflected energy and total energy can be called as albedo. The white surfaces like bright sand, ice and snow can exhibit high albedo ranges. The dark surfaces can absorb solar radiation and hence exhibits low albedo rates. Asphalt, ocean and dark green trees can exhibit low albedo range. The solar energy can hit the ocean surface resulting in the evaporation of water vapor. The energy carried by water vapor can be called latent heat. The radiant energy can encircle our earth system. This energy can flow via the atmosphere from one place to another. The migration or movement of water via the atmosphere, ocean, surface of the earth can be defined as the hydrological cycle. This cycle can participate in carrying energy via the atmosphere. Solar energy can yield vapor on hitting the ocean surface. The plants have an inherent capacity to transfer water back to the atmosphere via the transpiration process. The energy carried by the vapor can be conserved whilst it moves via the atmosphere. When the water vapor is subjected to the condensation process, there will be the formation of ice crystals.

6.3 GLOBAL WARMING

The processes and components of the climate system can be integrated to produce a three-dimensional complex. The feedbacks can be classified as an important component in any climate system. They control the climate sensitivity. If the carbon dioxide concentration increases, due to this temperature rise and there will snow melting. Due to this, the darker surfaces will get increased resulting in more absorption capacity or decreased albedo. This can be called as positive feedback. Due to this the climate sensitivity is high to the changes or alterations occurring in the systems. These fluctuations can be attributed to the energy budget. The anthropogenic factors can affect climate systems.

6.4 CLIMATE MODELS

Every component and process should obey the laws of chemistry and physics. The principles governing the climate systems cannot be classified as novel and we are not making it complicated anymore. Any climate model can be classified as an interlinked process that may occur in a series. The climate systems can be based on physical laws. Most of these laws can be solved and it is according to the location where this model is being tested. This can be called a finite element. The laws of physics must be solved at a grid box, cell or point or any physical location. Most of these models can be constructed on a vertical dimension. It can be soil, ocean and atmosphere. This will yield a column. This can be considered as one dimension with respect to vertical. These columns can be classified as grids or regular grids. We can assume each location as a grid point. The grids are nothing but the points that make up a lattice. Most of these grids can be equally spaced. In some cases, they can be irregular. This can be due to the diverse arrangement of the points. Any model can have three dimensions named as two verticals and one horizontal. Every vertical location that can be present in a column can be called as a cell or a grid box.

We can consider each cell as a finite element. The models should define this finite element present in processes. We can assign a single finite value for a region. In a point or a box model, we may not have dimensions. If we assume a single column, then there will be one dimension in the vertical. In the case of a three-dimensional model or 3D model, we can have one vertical dimension with two horizontal dimensions. In case, we are assuming a sphere, we can say that there is a 3D grid. When considering the resolution of a specific model, it is essential to define the size of the horizontal box. We need to consider the space that can be present in between two imaginary boxes. So, if we consider one latitude degree, then we can say that this box covers 110 kilometers. The longitudinal lines and longitudinal lines can produce a global grid. At any given latitude, we can find the same number of boxes. The components in the climate systems can be vertical and horizontal. It is essential to define the processes in two directions if possible. The horizontal processes comprise weather systems that are impregnated in the atmosphere, wind forces acting on the surface of the oceans and river flow. Most of the features that participate in climate systems are considered as horizontal in nature. The elevation and vegetation of the land can change but the ocean surface can be classified as a uniform surface. Some of the vertical processes are sinking of water in the ocean, rising of water in the ocean, vertical migration of air, percolation of water into the soil layers. Both buoyancy and gravity will affect these vertical processes. Every box should be capable of estimating the physical state of any system. The air and temperature in the box can be classified as essential variables. These variables can define the components in the box of our interest. Some of the physical properties are vegetation cover box over the land surface, soil moisture, salt content in the ocean box, ozone and water molecules present in the air box. Parameterization can be observed in the climate models. It is important to ensure that the small-scale natural process may need approximations. The energy mass should be conserved according to the conservation laws.

Chapter

7

Agrometeorology and Crops

CONTENTS

7.1 INTRODUCTION

The science that deals with the concepts of meteorology in relation to agriculture can be called as agrometeorology (Figure 7.1). We can use this science to judiciously use land resources and enhance food production rates. It also helps us in controlling us from damaging the land resources (Stern, 2010). This science can integrate hydrological, climatological and meteorological conditions with agriculture. The scope of this subject encompasses humans, animals, plants and other living organisms along with their interactions with the atmosphere.

Figure 7.1: Agrometeorology.

Source: http://www.mtx.it/en/applications/agrometeorology.html

It is to be considered that agrometeorology is an interdisciplinary subject. It involves soil sciences and atmospheric sciences too. All the perils related to the natural resources can have a meteorological component in their processes. Farm operations, forest fires, frost, the incidence of diseases and pests, soil erosion, air pollution and soil pollution can be linked with this meteorology component. Agrometeorology can help us in deriving solutions to use climate situations and avoid the risks associated with the climate associated perils. The effect of agrometeorology can be tactical and strategic in nature. Weather can affect the world agriculture in the form of storms, hail, frost, rains, flash floods and drought. If agrometeorology products are used, there will be minimal damage to the crops and sustainability will be managed (Rijks et al., 2000). Plant diseases, insects and parasites can be affected by the weather and climate situations (Mavi, 2004). With the data available, we can easily perform certain short-term adjustments in the agricultural fields. This will help us to get better yields. Advances in meteorological techniques can help us to get reliable data. The agrometeorological models can aid us in developing land-use planning that can help us in sustainable development. We can select the crop type that can work better in a given set of climatic conditions. We can also calculate the farming rates and even plan to adopt the latest machinery for the development of the yields. Technologies were developed to relate the environmental conditions with the biological responses. These responses can be quantified using novel technologies. These sophisticated facilities can help us in understanding the animals and plants effectively and we can relate their responses to the weather conditions. The application of the simulation models has helped many administrators to understand the role of the weather models in framing decision systems (Hoogenboom, 2000). Increase in the computation capacity has drastically reduced the costs incurred in research and development. Many efforts were put into scaling up the models from the leaf scale to the global levels and with this, we can get the data of the entire globe in a day or two.

Remote sensing can extract information from physical objects without coming into contact with them. We can measure and even characterize the objects on the surface of the earth. This can be done with electromagnetic energy. The agriculture sector can be classified as sensitive and most vulnerable as far as the national economies are concerned. All the nations are interested in providing specialized forecasts for agriculture. These countries develop agroclimatology services to the farmers and the general public. There is a need for a common database that can be available to the researchers and the general public to know the anticipated effects of the weather patterns. It is essential to procure both agrometeorological data and meteorological database so that it can be used in forest production and agricultural production. It can be important to use archival data to forecast and nowcast. The agrometeorological models can help us in framing policies, crop management and in crop research studies.

It is important to know the radiation and transmission effects of leaves against sunlight radiation. The radiation process can be highest at the green light. The transmission process can peak at the infrared region. The green color of the plants can be due to high reflectivity. The infrared reflection from the plants can protect them from uneven heating. Most of the plant canopy can absorb the incident radiation and 15% of this radiation will get reflected whilst 10% is transmitted. The plants absorb different wavelengths and it differs with the components of the leaves. If the epidermis is transparent, the sunlight can enter the mesophyll regions. This region contains palisade and spongy parenchyma. The former comprises cylindrical cells and they are closely packed. The spongy parenchyma comprises irregular cells and there will be air spaces. Both of these cells have chlorophyll. They can reflect the green wavelength. The blue and red radiation can be absorbed whilst photosynthesis (Sabins, 2004).

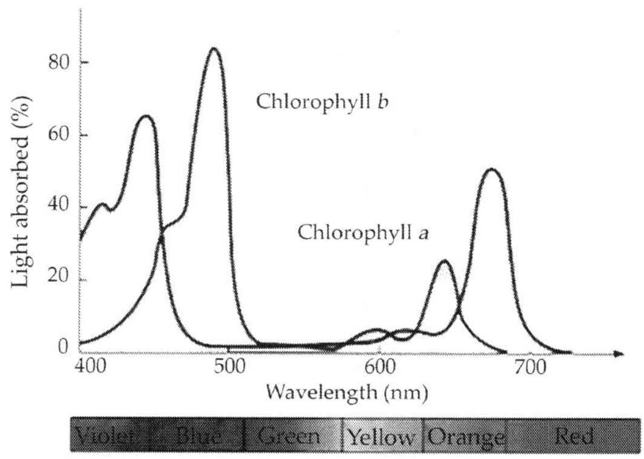

Figure 7.2: Chlorophyll reflectance.

Source: https://www2.mcdaniel.edu/Biology/botf99/photo/p3igments.html

Chlorophyll (Figure 7.2) can absorb blue and red wavelengths. The IR energy can get into spongy parenchyma. At this region, energy will be reflected and scattered by the boundaries between the air spaces and cell walls. The near-IR reflectance exhibited by the leaves is not due to the presence of chlorophyll. This can be due to internal cell morphology or the internal cell structure of the leaves. The IR radiation can be absorbed by the plants if the wavelength is >3 micrometers. The leaves absorb far-IR wavelengths. The near-IR radiation cannot be absorbed by the leaves because it can enhance the internal temperatures of the leaf. (Figure 7.3)

The radiation quality can affect all the growth phases of the plant. The flowering stage, germination and elongation stages will be affected by the radiation quality. In the case of long-day plants, red light (0.66μm) can lower the flowering process. In the case of apples, red light can help in the production of red color. If the seeds were subjected to short, blue and green wavelengths, then their germination will be inhibited. We can induce seed gemination by passing red wavelength light. There will be an elongation of the stem in the far-red wavelengths. It is to be noted that the red portion of the spectrum lowers stem elongation. The visible wavelengths can affect the shoot orientation. This can be called as phototropism (Kato et al., 2019). We can observe positive phototropism. If the intensity of the light is increased, then plants exhibit negative phototropism. The gamma and UV radiation can have less effect on plant growth. We need to know that these portions of radiation can produce biological effects (Gatti, 2009). This portion of radiation can reduce the incidence of diseases and kill microbes.

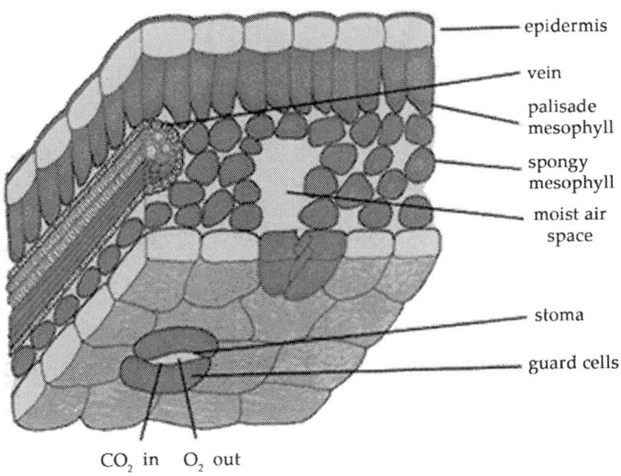

Figure 7.3: Leaf anatomy.

Source: https://projects.ncsu.edu/project/bio183de/Lab/photosynthesis_lab/photosynthesis2H.html

The solar spectrum can be classified into bands according to the physical responses exhibited by the plants. If the wavelengths are <0.28μm, then we can assume that there will be a strong germicidal action. They can damage eyesight and it should be known that this radiation is not prevalent on the surface of the earth. If the wavelength ranges from 0.31 to 0.28μm, we can expect germicidal action. It is obvious that the wavelengths less than 0.29μm may not reach the surface of the earth. If the wavelength is between 0.40 -0.315μm, we can expect fluorescence in plants. We can assume firm formative effects and strong photosynthetic activity if the wavelength between 0.610-0.510µm is to be used. If the wavelength lies between 0.7-0.6 µm, we can expect strong periodic activity and this can be attributed to the absorption of the light by chlorophyll pigments. In case, the light with a wavelength between 1-7µm, we can expect plant elongation.

It is to be known that there are many aspects of solar radiation that are to be studied whilst knowing the biological responses. The amount of radiation that can reach a unit surface area at a given time can be called as the intensity of radiation and this can be considered an important aspect. The spectral distribution that can regulate photosynthesis can be treated as important. The distribution of radiation according to time can be considered as an important aspect in determining the photoperiodism. It is essential to quantify the spectral distribution and intensity above the crop canopy. This is important to understand the microclimate and photosynthesis process. The transpiration rates can be affected by the radiation intensity with time. Radiation can be used to produce dry matter and it is based on photosynthetically active radiation or PAR. (Figure 7.4)

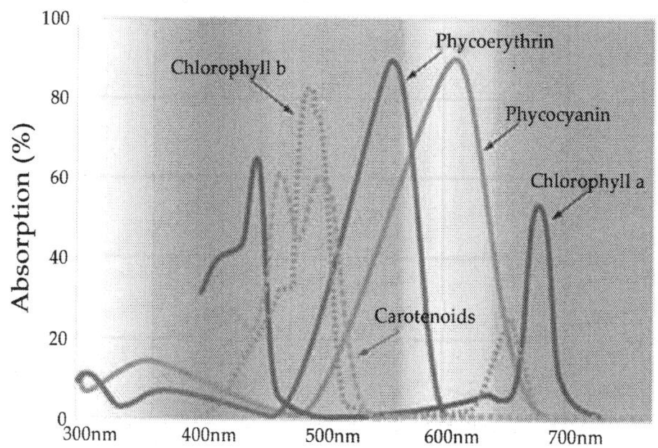

Figure 7.4: Photosynthetically active radiation.

Source: https://www.blackdogled.com/blog/photosynthetic-active-radiation-par/

The radiation distribution can be determined by many factors. It will be based on the sun's angle, height and density of plants, inclination and arrangement of leaf and leaf transmissibility (Li et al., 2005). The transmissibility can be high in the case of young leaves. The leaf transmissibility can be linked with the chlorophyll concentration. If the chlorophyll concentration increases, then the transmissibility rates will come down. If the height of the plant increases, then the light intercepted by the canopies will get enhanced. This will be varied according to the day timings. When the solar radiation reaches the canopy, there will be transformation throughout its passage via the layers (Myers, 2013). The IR and red radiation can get increased with every transmission and reflection. The rate of solar radiation will get fluctuated with the sun's angle. It is to be noted that the spectrum on the crops will be high in the noon. IR penetration can be high in the evening times prior to sunset. The visible portion of the spectrum can be classified as PAR or photosynthetically active radiation. The unit for global radiation is W/m^2 whereas the unit for PAR is μE/m^{2}s. The number of photons that can be present in the PAR or band can be called as PPFD or photosynthetic photon flux density. (Figure 7.5)

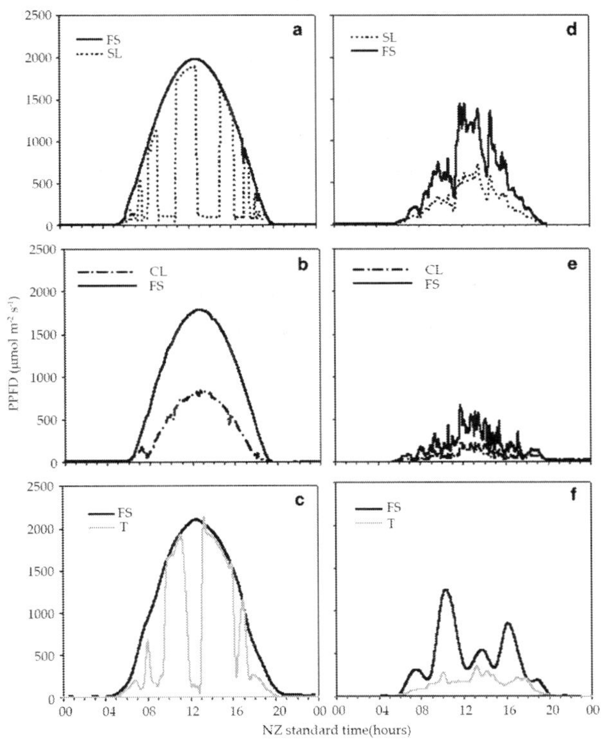

Figure 7.5: Photosynthetic photon flux density.

Source: https://www.researchgate.net/figure/Photosynthetic-photon-flux-density-PPFD-measured-in-full-sunlight-FS-under-wooden_fig8_225691457

The unit for PPFD is mol/m^{2}s. PAR can be measured as a ratio derived from solar radiation. It is essential that we need to calibrate PAR according to the regions. This is due to the fact that PAR can be dependent on other local geographic and climatic factors. Some of the factors are the diurnal pattern of radiation, length of day and cloudiness rate.

The interception of PAR in the wheat crops was studied by several researchers. Two wheat varieties were cultivated for a year and the albedo, net radiation, PAR and solar radiation interception were measured. It was observed that post 100 days of sowing, there was maximum PAR and solar radiation interception. If nutrients and water are not needed to limit the plant growth rates, it is important to know that biomass production is controlled by radiation. Solar radiation can be considered as the primary heat source available to the biosphere. Temperature can be an important aspect of organic life. This can regulate both the physical and chemical components of the landscape. Temperature can also control the diffusion rates of liquids and gases in the plants. Most of the solubility rates can be based on temperature rates. The temperature has a critical role in the growth of plants and it exhibits diverse geographical distributions. The temperature of the soil can be considered as an important aspect in the growth and distribution of the plants.

There are several factors that affect the temperature of the soils. Aspect and slope can be considered as important. In Southern Hemisphere, the north-facing slope can be warmer than the south-facing slope and this will be reversed in the case of the NorthernHemisphere. It is important to know that the temperature of the soil is greater than the air temperature. Tillage can be important when studying temperature patterns. The heat flow between the subsoil and surface can be reduced by tillage. Due to tillage, the topsoil will be loosened and there will be the production of mulch. The mulched surface can have more exposed surface area. The link between the layers below will be cleaved and hence increased exposed surface. The cultivated soils can exhibit high temperatures than the uncultivated soils. Soil texture can be considered as an important factor. The sandy soils can be easily heated and can get cooled. This is not the case with clay soils. The clayey soils can store heat for longer periods. The organic matter can lower the thermal conductivity and heat capacity of the soil. The organic matter can enhance the water holding capacity of the soils. Due to the dark color of the soils, we can expect increased radiation absorptivity.

The temperature of the soils can affect the seed germination and activity of the rhizosphere. This can also affect plant growth rates and the incidence of plant diseases. Most of the living tissues can be damaged if exposed to increased temperatures. The stem and roots will get damaged if exposed to the high soil temperatures. It should be considered that the low temperatures might affect the plant growth rates. Extremely low temperatures can inhibit the nutrient absorption capability of the plants. The effect of soil temperatures on a plant will be higher than the air temperatures. It should be known that most of the rice varieties may not start to grow if the temperatures are less than 11^{0}C (Kwon et al., 2004).

If the soil temperature is above 21^{0}C and below 12^{0}C, then the germination of soybean, maize and sunflower will be halted (Hernandez et al., 2020). High soil temperatures can damage the potato tubers. Post germination, the temperature can be considered as an important aspect in the vegetative phase of the plant species. The uptake of iron by the plants will be affected by the temperature fluctuations. It is important to know that the day time temperature is important than the night time temperatures.

Rainfall or precipitation helps in the production of food all over the world. More than half of the food produced worldwide is dependent on rainfall to this day. The rest of the world is based on irrigation. Most of the crops are affected by the disturbed rainfall patterns annually. Most of the global waters are being used for feeding purposes. Irrigation can be considered as an essential option to get sustainable crop production. Most of the investments in the food production sectors are based on irrigation only. Less investments are being made in the production of rainfed crops. Due to unprecedented irrigation, most of the land water resources are being endangered and it is bringing us all the natural calamities. Since the water resources are declining, it is important that water must be saved for the present and future. It is essential to plan and manage the water resources properly as this will rejuvenate the groundwater and surface water resources. The temperature extremes and inadequate water supply can be considered as major risks in crop production. There are many studies that were designed to identify the rainfall patterns. This will be used in the frequency of irrigation, the period of irrigation, harvesting date and sowing dates. These studies can be made with the help of statistical analysis. We can use historical data for that matter.

Plants absorb water that will be lost to the atmosphere via transpiration. This process can be classified into lenticular transpiration (Figure 7.6), cuticular transpiration (Figure 7.7) and stomatal transpiration (Figure 7.8). The transpiration via stomata can be defined as stomatal transpiration. This type of transpiration essentially occurs in the leaves. In the leaf external anatomy, we can observe a layer over the epidermal cells. This is called as the cuticle. In this cuticle, there will be pores or gaps. Water can get escaped from these gaps. This can be called as cuticular transpiration. There will be gaps or pores in the stems and roots. Water can get escaped from these gaps or pores. This process can be called as lenticular transpiration. The transpiration rates can be based on crop type and meteorological conditions. Stomata generally get closed in the night and get opened in the presence of light. Stomata can get opened in the day times. If there is less humidity, we can expect high transpiration rates. If the difference between the leaf humidity and atmospheric humidity increases, there will be enhanced transpiration. The vapor pressure and humidity can be considered as a function of temperature.

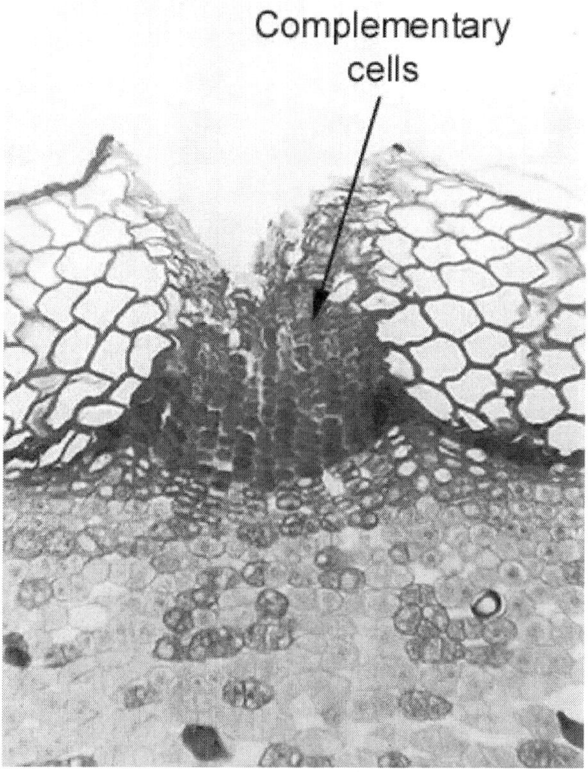

Figure 7.6: Lenticular transpiration.

Source: https://www.quora.com/What-is-lenticular-transpiration-What-are-some-examples

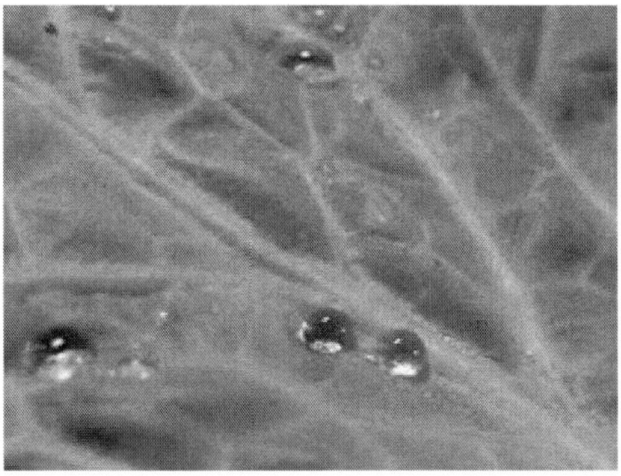

Figure 7.7: Cuticular transpiration.

Source: https://en.wikipedia.org/wiki/Plant_cuticle

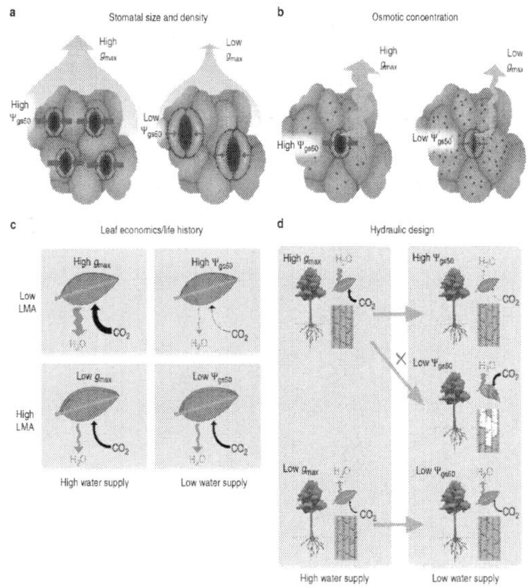

Figure 7.8: Stomata transpiration.

Source: https://www.nature.com/articles/s41467-019-11006-1

If the temperature decreases, there will be increased vapor pressure. This will lower the saturation deficiency. If there are windy weather conditions, most of the dry air enclosing the plants will be transformed with fresh air. This will enhance the transcription rates. The root to shoot ratio should be high, then it is possible for the plant to absorb more water and so reduced transpiration. The physiology and morphology of the leaves can define the transpiration rates. In the case of large leaves, we can expect high transpiration rates. If the cuticle is thick, then we can assume that there will be less transpiration. The epidermal hairs on the surface of the leaves can minimize the transpiration rates. There will be diffusive resistance in the plants whilst transpiration and this property will demarcate the transpiration in plants from evaporation. The evaporation rate on the land surface can be guided by certain processes. It is based on the water movement in the soil strata. Water can move from the surface of the soil to the root system. The water molecules will enter the roots and can reach several plant tissues. Water can get transformed into vapor at stomata. These processes can be called as evapotranspiration or ET. It is the combined effect of evaporation and transpiration. The evaporation dominantly occurs in the soil. Crop growth can depend on ET and meteorological factors. This can be called as potential ET. This can be classified as the upper limit of ET (Figure 7.9). This is based on the evaporating tendency of the atmosphere. The stomatal behavior and aerodynamic properties of the plants can alter the effect of ET. We can define PET accordingly.

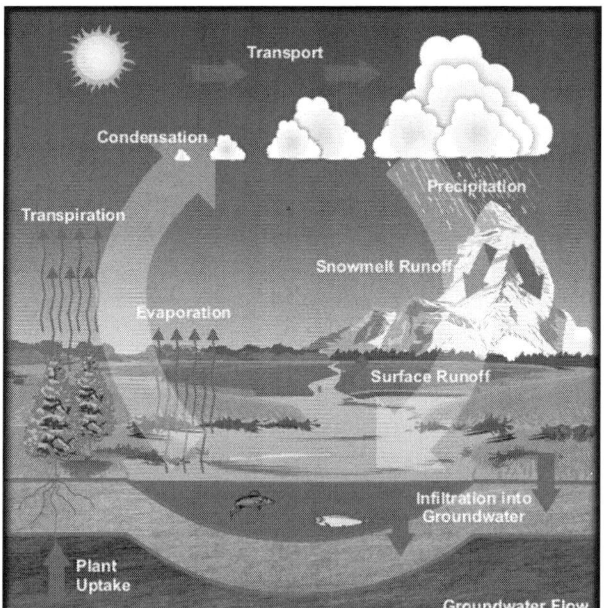

Figure 7.9: Evapotranspiration.

Source: http://w3.sista.arizona.edu/~clayton/evap-web/evap.html

7.2 DROUGHT MONITORING

Drought can be classified as a serious climate hazard that can shatter the food resources. This can cause environmental degradation, economic losses and a lot of physical suffering. Drought can be classified as a complex phenomenon that can be hard to define. In some cases, dry spells can get transformed into severe drought and this will get evolved as exceptional drought. (Figure 7.10)

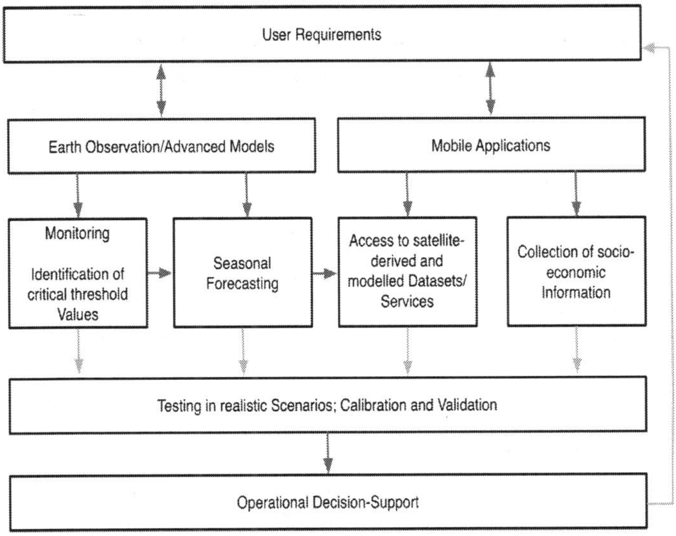

Figure 7.10: Drought monitoring.

Source: https://un-spider.org/links-and-resources/data-sources/daotm-drought

The transformation of dry spells into a regular or severe drought can be a slow process leaving long-lasting ill effects that the local and global populations might endure. Relatively speaking, the effect of drought will be slower than other disasters. The miseries and the costs incurred whilst drought and after can be higher than any other disasters. Due to drought conditions, there will be a water shortage in the soil layers. This will lead to reduced wildlife stock and hence most of the forest resources will be damaged. There will be a sequence of indirect effects of drought conditions. This can induce population migration, famine, disease and malnutrition (Stehlik et al., 2016). The costs that are linked with drought can affect the environmental, social and economic factors. Some of the economic costs incurred can affect timber production, fishery, livestock, dairy and crops. The social costs can include costs that can damage the urban ecosystem and rural ecosystems. This unrest can be due to population migration, social unrest, lowered quality of life, enhanced poverty levels along with garbage and water associated problems. There will be several conflicts that can be associated with drought conditions.

Meteorological drought can be defined as an abnormal lowering of the rainfall rate over a specified time period. Agricultural drought can occur if the moisture levels in the soil layers are reduced to a minimum at a specific period. The hydrological drought can be defined as the deficiencies that can appear in the subsurface and surface waters. Most of the drought conditions can occur due to water deficiency in space and time. We can observe deficiencies in soil moisture, ground aquifers, stream flow and precipitation patterns. Based on the area that drought affects, we can classify drought as continental drought, small-area drought and point drought. It must be known that the small-area drought and point droughts can occur frequently.

There are several methods used in the assessment of drought. Some of these methods can be based on qualitative observations rather than quantitative data. It is important to conduct field surveys and it must be based on scientific criteria. The field surveys should be planned such that the error rates are at a minimum. It can be mentioned that there is no universal method for assessing drought conditions as the assessment strategies can change according to the local conditions. Each country will follow its own method to analyze the quantitative and qualitative data of its land area. When assessing drought, it is important to record the normal precipitation percent of that place. This will help us know the exact nature of drought that occurs in the landscapes. This can be derived from dividing precipitation values with normal precipitation values. In general, we can consider 30 years of mean data and then we can multiply with 100. The normal percentage value can be measured for diverse time scales. We can get better to analyze the normal rainfall percentage over a single region. The conditions that determine drought can vary with location. Any area with more than a 20% chance of drought occurrence can be called as a drought-prone area. If this rate is at 40%, then we can define it as a chronic drought-prone area. If the rate is at 75%, we can call that year as a drought year. In order to minimize the percent of normal values, a system based on ranking rainfall values was devised (Gibbs et al., 1967). The rainfall data abstracted from long-term records can be classified into sections. Each section represents 10% of the distribution. This was named as a decile.

The dependable rains can be called as DR. This is nothing but a rainfall occurring every 5 years and this was observed in African countries (Le Houerou et al., 2009). These rains can be used in the planning strategies of agriculture and related areas. According to many researchers, this index lost credibility. In order to reduce the error rates whist measuring the precipitation deviations, the National Rainfall index was established. The rainfall index system can compare with the data obtained from several years. This index can be free from the absolute rainfall values. The utility of rainfall can be less if we want to calculate drought conditions by taking into consideration all the impacts.

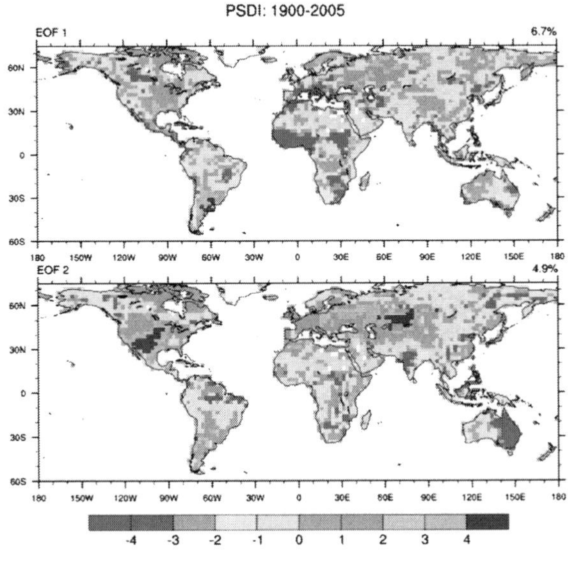

Figure 7.11: Palmer drought severity index.

Source: https://climatedataguide.ucar.edu/climate-data/palmer-drought-severity-index-pdsi

There is another index called as Palmer drought severity index (Figure 7.11). It can calculate the deviations in the moisture levels (Siska, 2017). This index can be based on the meteorological drought index. It uses temperature data and precipitation data obtained from the ground stations (Steven et al., 2009). Palmer Index is used in the United States. This index can be used to evaluate the impacts caused due to soil moisture conditions. We can classify this index as an effective drought monitoring tool. This can be employed as an important indicator. This can be used in making contingency plans. This index will allow us to represent both spatial and temporal representations effectively. We can relate current weather conditions with the archival data. We can accurately estimate and measure the weather variations in a region. Weather can be considered as a determining factor that can define the periodic abundance and geographical distribution of the animals, parasites and insect pests of crops. The weather pattern can regulate the insect outbreaks, migration, phenology and development rates (Pedgley, 2009).

Weather can affect all the resources at various time scales. The time-lagged influence can happen because of the existing weather situations and past weather conditions. It is to be noted that wind, humidity and temperature play an important role in determining the weather conditions. These three can be classified as important weather elements. Every insect species can be sensitive to temperature fluctuations. There is a temperature range that defines the life cycle of the insect. This can be called a tolerable zone and it means that insects survive at this temperature without damaging their tissues (Atkins, 1988). Every physiological function in the insect life cycle is controlled by the temperature range. Once, the temperature reaches its threshold value, most of the vital functions of the insect will be halted or lowered. This will result in the death of the tissues. When the tolerable zone hits its extreme, the insect populations will start to die. The values of actually tolerable zones may vary according to the species. The exposure time can be considered as important as the duration of exposure can explain the survival rates of the insects. When subjected to high temperatures, the survival rates will be lowered (Kaakeh et al., 1997).

Some insects can grow a function faster if subjected to high temperatures. They can easily reproduce and can disperse their progeny (Drake, 1989). The velocity of the wind, vertical temperature profiles and other factors can affect the movement of the insects and hence the migration rates will be lowered. The temperature values can define the duration of the flight, height of the flight and time of the flight. The behavior of the insects will be regulated by the temperature variations. The moisture content can be an important factor as far as the insects are concerned. Most of the parameters that control moisture can affect insects.

7.3 REMOTE SENSING

The spatial information can be important in many aspects of crop planning and management. It is nothing but geographically located data. This data can provide us information about objects or phenomena present on the earth's surface. This data can be retrieved from many sources like weather station reports, field-recorded surveys, aerial photographs and satellite imagery. The essential property of this data is that we can get the information of the objects that are closer to each other though they are not related. Spatial information can be classified as a wide area that can give us information about various aspects like climate data, weather data, administrative boundaries and roads ,etc. Some of the data can be collected from the ground stations and others from the satellite imagery and so on. It is important to consider that remote sensing is an essential tool that can provide data pertaining to almost all the objects and events that occur over the surface of the earth. Mostly, we can consider points, line and polygons or even areas as tools to demarcate the spatial information. The soil profile characteristics from a point can be considered as point data. In the same way, an irrigation channel can be classified as line data. Water catchment can be classified as polygon data. The spatial data can be stored, analyzed and mapped using computer-aided geographic information systems. We can use a global positioning system in this case to get accurate data about the objects. Remote sensing can be used to know the characteristics of the phenomena and objects that are present on the earth's surface. We can interpret the data obtained using GIS software as and when required. (Figure 7.12)

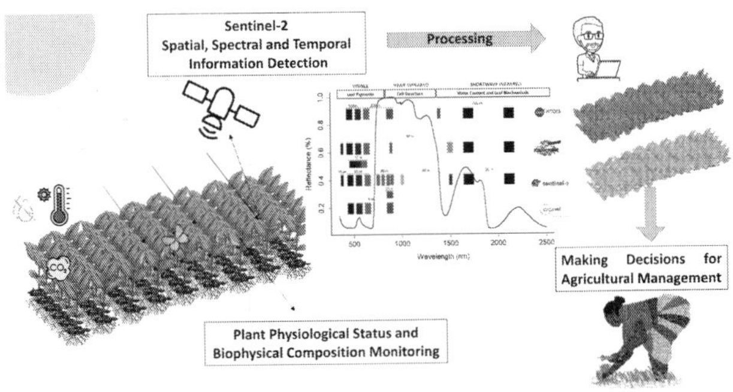

Remote sensing for agriculture monitoring: Sentinel-2 features and precision agriculture

Figure 7.12: Remote sensing and agriculture.

Source: https://www.mdpi.com/2073-4395/10/5/641

We can estimate the vegetation cover extent on the surface of the earth and this gives us an important estimate of the fodder available to animals. This data will help us understand the soil resources and we can make plans to manage the soil resources. We can use crop models, pasture models, field measurements and remote sensing to estimate the cover. Remote sensing technique is being used by several developed countries to achieve the best results in agriculture. In general, we can use the Normalized Difference Vegetation Index or NDVI for this cause. NDVI can be calculated using the NIR and Red wavelengths of the radiation reflected from the objects.

The first attempts in the usage of plant and animal simulations started in the 1970s. The models used were simple and some are complex. The complex models can take many inputs and so supplying data to these models as inputs can be a hectic task to be considered. Most of the complex models are restricted to certain zones where all the tools to obtain data are in place (Jorgensen, 1978). Many analytical procedures pertaining to both physical and biological phenomena were linked with the models (Decker, 2008). The crop growth models can help us in solving many questions in crop research. These models can help us relating environmental factors with physiology and genetics. We can correlate in-season and preseason management of crops using this technique. These models can help the policymakers in estimating the soil erosion, climate change effects, leaching and other harmful effects of pesticide use, irrigation, fertilization and other cultural practices. These models have yielded many economic advantages over conventional methods.

Any model can be defined as method that can be used to represent the conception of a particular system. Most of the models can explain the functional forms of any real object existing on the surface of the earth. These models can aid us to simulate situations that may or may not surface in the real life. We can simply explain these models as mimicry. These models can be classified as dynamic simulation models and statistical models. Statistical models mostly rely on techniques like regression or correlation pertaining to the environmental variables and plants (Norman, 2015). These statistical models can be classified as crop-yield weather models. These models can be used to estimate the production yields over vast regions. The regression coefficients may be linked with the processes involved in determining crop growth. It is always important to create regression equations so that the model can have wide-spread application. It is essential to use or perform several studies so that these regression equations can be produced without errors. One of the important advantages of these models is that it can be used in weather-crop models readily. This will help us in easy forecasting and management. These models can be used to recognize the crop failures and crop success rates using existing data and archival data. These regression models can be considered as the best choice as they can be simple and straightforward. They can give us a link between the environmental factors and yields associated with these factors. These models might not be accurate in all cases. These models can be used in the prediction of the crop yields for a single crop that can extend over a wider area. The regression models can take into consideration all the insect-disease problems, agronomic practices and soil types. These factors can be combined to get an accurate effect.

Any dynamic model can be linked with the time periods with respective outputs. In this model, it is important to characterize the processes. In order to do it, it is important that we know the state variables. The state variables are those variables that can describe the state of a particular system at a certain period. These models can predict fluctuations that may occur in the crop growth and we consider time as an essential function with relation to biogenic values generated (Hume et al., 1995). Here, simulation can be described as a model that can near-accurately act as same as a crop in the field. In this, we can enhance the simulations using roots, stems, growing leaves and germinating mechanisms. This process can be constructed dynamically over a specific time period. There are different types of dynamic models. They are preliminary models, summary models and comprehensive models (Penning de Vries et al., 2020). In the case of preliminary models, we can observe structure and data. These models can be classified as simple. Their simplicity can be attributed to the exploratory levels. In the case of a comprehensive model, much of the knowledge gained through various sources will be examined and analyzed. These models can predict yields and the results can be the same as the real events. They are better than the regression models. If we need a more accurate dynamic model, we need to input more data from reliable resources. It can be important to establish an expert team to run comprehensive and dynamic simulation models. Most of the soil scientists, agronomists and plant physiologists can take part in this modeling process. With these models, it is obvious that we can establish an appropriate relationship between the plant growth parameters and environmental factors.

It is always important to verify and validate the model so that it can be used with various variables. In the verification process, we can find out the correctness and truthfulness of the model and the outputs it generates. The verification process can establish functional relationships that can be appropriate and free from errors. In the case of variations in the expected outputs, we can change the model parameters and the programs as per the requirements. Validation is yet another important aspect of the modeling process. It must be conducted by performing independent observations at various places so that the validity of the models can be deemed as fit for all. Any model can be classified as valid only if the values derived from the model falls under the confidence band that is previously projected.

The simulation models of the crops can be important to know the crop management strategies and application requirements as and when needed. We can use crop models to understand the actual situations that might occur in the future. We can get measure the harvestable products, biomass yields, tuber yields, root yields, fruit and grain yields. There are several simulation models that can use the genetic characteristics of the crops. These constants or coefficients can be evaluated to know the sensitivity of the models. With these conventional techniques, it is not possible to get data on all the physiological processes. Turgor pressure can be considered as an important aspect in the expansion of the leaves. The main problem with these models is that this turgor pressure cannot be measured accurately with these models. We can measure the turgor pressures with the aid of dynamic simulation models. It is to be known that these models can be comprehensive and so we can test the efficiency of these models on several growth factors. We can even stop considering hypotheses that are unrealistic in nature. We can probe the physiology of the plants that are otherwise impossible to perceive. It is important to know that experimentation and modeling can go hand in hand with symbiotic associations. This concept can be essential to know the crop physiology completely. It is important to know the crop-rotation analysis and its sequence properly. We can analyze crop rotations. With these simulations in place, we can analyze multi-year data. It can be important to simulate carbon, nitrogen and water present in the soil and on the surface. The important target is to know the long-term fluctuations that may vary with time. There are many models that are being developed to study the nutrient dynamics of the soil with reference to nitrogen and organic matter.

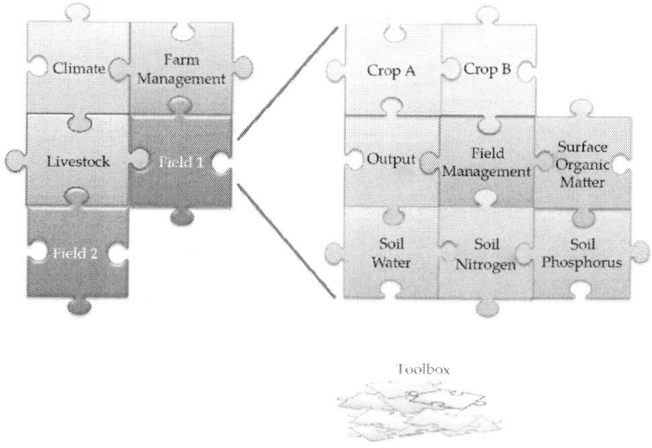

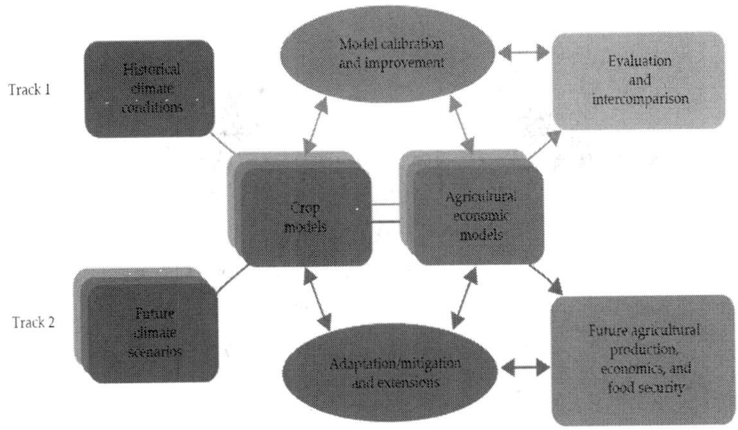

Figure 7.13: Agricultural Production System Simulation Model.

Source: https://www.sciencedirect.com/topics/agricultural-and-biological-sciences/crop-simulation-model

We can use the Agricultural Production System Simulation Model (Figure 7.13) or APSIM to know the crop sequences and evaluate them. These models can be important in comparing the crop management strategies and correct them is not satisfied with the predefined limits. Most of the management decisions can be controlled by these models. The management inputs and biological inputs can be merged with several other factors so that the actual risks can be forecasted (Thornton et al., 1993). The crop models can be run before the crop growing season so that the crop growth patterns can be associated with the models. There will be some uncertainties with this modeling process. The uncertainty in the weather parameters should be normalized so that the error rates can be minimized. We can use historical multiyear data and feed it into the model so that we can forecast the current and future scenarios with utmost confidence. We can use the generated data along with the historical data as most of the data will be reliable.

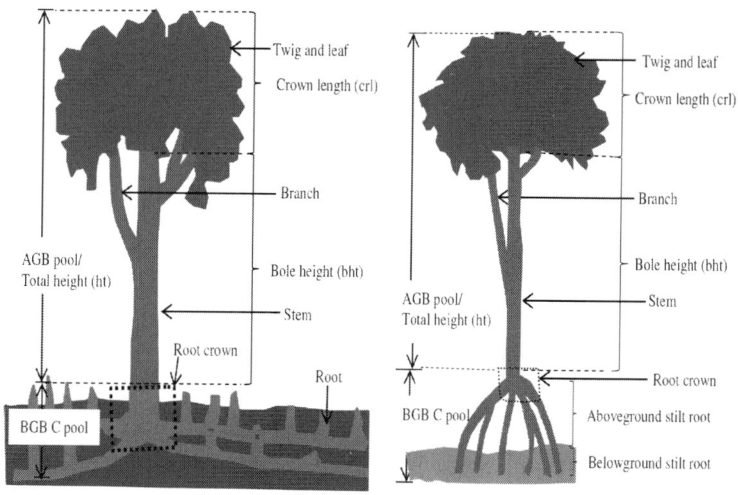

Figure 7.14: BIOMASS tree model.

Source: https://link.springer.com/article/10.1007/s13595-015-0524-3

The BIOMASS tree model (Figure 7.14) can give us information about the growth, dry matter and crop yields (McMurtrie et al., 1989). The CenW tree model can give us information about nitrogen cycling, water use, carbon gain due to photosynthesis (Vose, 1982). APSIM or Agricultural Production Systems Simulator can help us in deriving data pertaining to interspecies competition, crop rotations, nitrogen balances, water and carbon dynamics (Munns et al., 1982). SIMTAG or simulation model for Triticum aestivum can help us to get the data about the grain yields, plant dry matter and other development phases (Kim et al, 2010). The wheat model can aid us in generating data pertaining to grain yield, grain number, crop biomass, leaf area index, crop growth rates and soil moisture content (Wang et al., 2012). QBAR or barley model can be used to estimate the grain yields, biomass production, leaf area, phenology and soil water balance (Hook, 1982). AUSCANE or sugarcane model can be used to know the sugar concentration of the cane and yield of the cane stalks that can be millable (Jones et al., 2020).

CERCOT or cotton model can be used to estimate both growth and production rates of cotton and similar plants (Hook, 1982). OZCOT model was framed for cotton plants (Hearn, 1994). It can be used to estimate the nitrogen uptake, leaf area index, fruiting rate dynamics, yield components, evapotranspiration and water content of the soils. GRASP can be used as a pasture model (Sapaty et al., 2017). This model can be used to estimate the live weight gain, utilization rates, diet selection, animal intake, detachment and death, growth of pastures and status of soil water. PGAP model can be called as pasture growth and animal production model (Lovell et al., 2012). This model can be used to know the total pasture growth in a year, pasture senescence, growth of pasture, animal intake, growth index, dead matter and green matter. DYNAMOF or dynamic management of feed can be used for estimating the net farm income, staple strength, the diameter of the staple, length of the fibre, the weight of fleece, sales of sheep, lambs sold and born, growth of wool, pasture digestibility and availability of soil moisture (Bowman et al., 1964).

There are several decision support computer packages that can be useful in the farming systems. RAINMAN can be considered as an important computer package (Gabriels., 2019). This model can be used to investigate the historical record of several parameters. Some of the factors are sea-surface temperature, seasonal rainfall, daily rainfall records, monthly precipitation, etc. Met Access package can be used to analyze and summarize the weather events that can occur daily and even it considers long-term data (Donnelly et al., 1993). The output in this model can be in tabular or graphical formats. We can calculate the long-term probability of the weather patterns that occurred in the past and forecast the future weather patterns that may likely occur in space and time. HOWOFTEN can be an important model in decision support systems. It can be useful in determining the crop selection process. We can also estimate the likeliness of the floods. We can know if there is sufficient rainfall with respect to the soil profile (Pal et al., 2020). PLANTGRO can be used to define the land suitability and it also describes the growing seasons of particular crops. It can estimate the plant growth rates according to the climatic conditions and local soil conditions. It can explain the months that can be useful in the planting process (White et al., 1997).

HOWEET can be used in calculating the rate of fallow rainfall and it also determines the amount of water that soil can store. Nitrogen accumulation in the fallow lands can be calculated using this model. It can be used in defining the irrigation needs, plant density, crop type selection and extent of fallow and how long will it take to land to become fallow (Dimes et al., 1998). MUDAS model can be used in integrating the crops with live-stocks. It can be used to enhance profits. This model can categorize the local situation into the quantum of rainfall in the spring, the rainfall received post break, the time before the onset of rains and rainfall collected in the summer or rainfall events in summer. It can help us to allow if any adjustments are needed in planning. Some of the adjustments are linked with the pasture deferment, storage of grains, use of fertilizers, sheep weight, use of machinery and crop area (Kingwell et al., 2015). TACT model can measure the probability distributions of the wheat crops. This can be specific to a particular location. This can use empirical relationships. This model can predict the exact timing of the phenological circumstances (David et al., 2012). STIN or stress index can be used in wheat yield model. It can be based on wheat yields and soil moisture. The soil moisture values can be included at the sowing dates. We get crop yields and soil moistures as output. We can use these values to predict the wheat yields at national, state and other regional levels (Stephens et al., 1979). WHEATMAN and BARLEYPLAN are popular models that can be used to both compare and estimate the outcomes that may result in the farming seasons. This model can be used in the cultivation of chickpeas, barley and wheat crops. These models can help us in the crop selection process, fertilizer application and planting time. It can be used to compare the economic factors and production factors that can control the properties of the crop plants (Woodruff et al., 2001).

PYCAL can be considered as an important model. This can be called a potential yield calculator. This program can be used to estimate the seasonal rainfall with relevance to the previous rainfall records. This model can be used to estimate the soil water content in the croplands. This model can give us the yield estimates of oilseed crops, pulse and cereal crops (Raman et al,2020). SOWHAT can be used as a decision support tool for farmers who can cultivate wheat. This model can work better in the climate like the Mediterranean climate. This model was framed by using the historical data and it also considers low probability and yield estimates. This model can be useful in maintaining sowing conditions. This model can enhance wheat yields. Whopper Cropper is one of the models that can be used to design cropping system models. We can do seasonal climate forecasting. This can be used in the management of crops. This model can offer a graphical user interface (Miller et al., 2013). The SIRATAC model can be used in the management of pests in the decision support systems. This model can be applied to cotton farmers (Bange et al., 2002). The EntomoLOGIC model can be used in the management of pests. This can be used in decision management systems for cotton farmers. In this model, there are three separate modules. We can forecast and standardize the spray counts based on the insect pressure over the crops.

The CLIMEX model can be used to design pest control systems, especially the insects. This model can be used to correlate the locations and climate with crop production rates (Sridhar et al., 2016). Grazfeed is another model that can be used by the agricultural advisers and farmers whose main interest is crops and grazing. This model can evaluate the sheep and cattle production so that the products like milk, wool and meat can be estimated beforehand. This can help us understand the production capacity of the crops and cattle. The GrassGro model can be used to estimate both growth and production of cattle and sheep. We can use this model in the planning of the strategies to be followed in the production of the breeding stocks. This model can be used to estimate the margins that can exist in the enterprise. We can use this model to compare the stock rates that might be present in the cultivation process. We can know about the diverse plant characteristics in the process of grazing animals. We can also compare the production rates of pasture. Sheep can be used in the production of wool. We can use this model in the development of gross margins, stock sales, lambing extent and in framing the management plans.

Weather can be important in the production of crops and this can be classified as a variable entity. It is essential to know that the farmers or the growers have no control over the weather or its consequences. The variability of the climate can be seen in most of the seasons and even drought may result in some cases. This will generate high pressures on the resources of land and even vegetation. It can be known that the previous land use management strategies are no good for the present climate scenarios and they seldom are suitable. Due to this, the rate of farm production will be reduced making the farmer communities vulnerable to the aftereffects of climate.

Chapter

8 Climate Management and Legumes

CONTENTS

8.1 INTRODUCTION

The effect of climate change is high in the legumes like common vetch, grass pea, lupins, lentil, broad bean, chickpea and dry pea. These legume crops can be classified as cool season legume crops. These crops were also named cool season grain legume crops (FAOSTAT, 2020). All these legume plants can be classified as C_3 plants. (Figure 8.1)

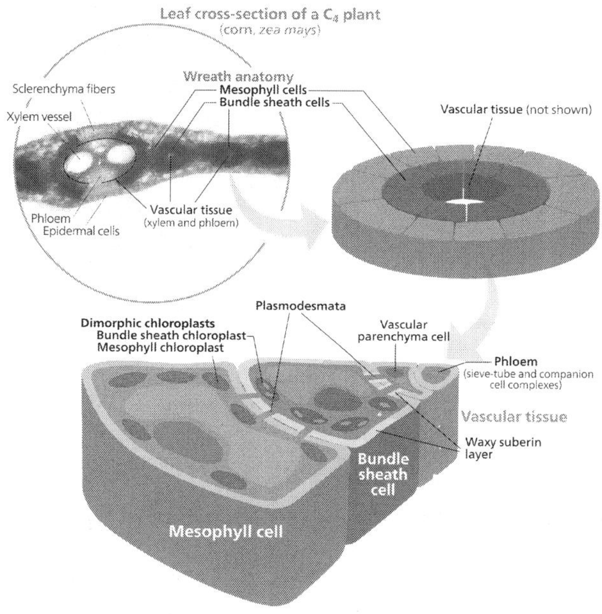

Figure 8.1: C_3 plants.

Source: https://legumelaboratory.wordpress.com/tag/c3-photosynthesis/

These plants can have a symbiotic association with rhizobacteria. The root nodules present in these plants can help in the colonization of these bacterial colonies. They can fix atmospheric nitrogen (Andrews et al., 2019). This additional functionality of these legumes can help them adapt to the nitrogen deficit soils. The popular cool season grain legume crops are lentil, broad bean, chickpea and dry pea. (Figure 8.2)

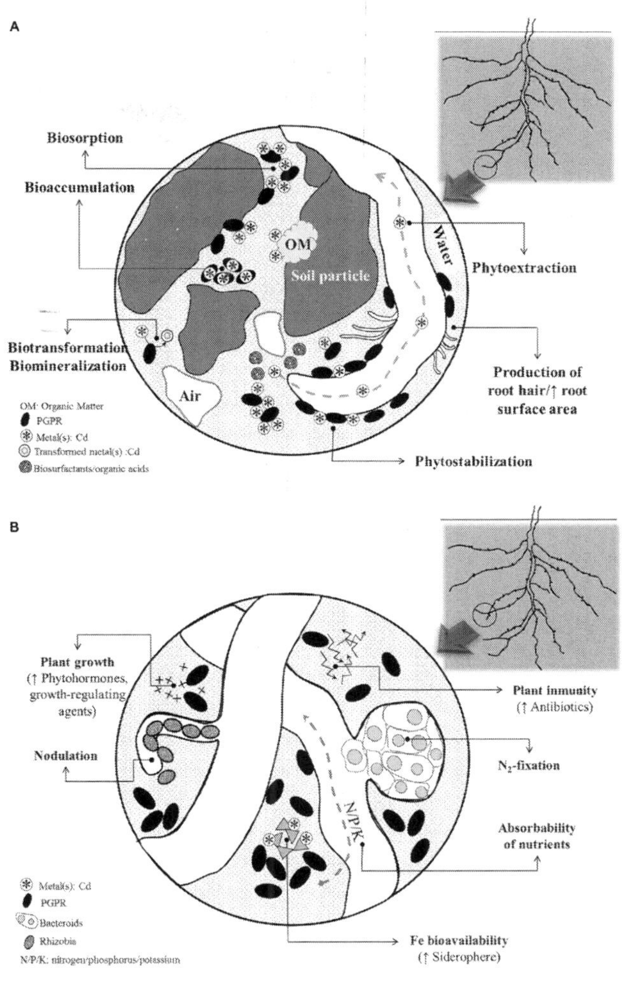

Figure 8.2: Nitrogen fixation.

Source: https://www.frontiersin.org/articles/10.3389/fpls.2015.00081/full

In most of the developed countries, these legume crops can be classified as minor crops. These crops can be used with other crops in the crop rotation process (Zohry et al., 2018). These crops play a vital role in the feed economy and food as they have protein content and they can grow in nitrogen-deficient soils. These crops can be used to regulate diseases, pests and weeds. This is the main reason behind the cereal cropping. Cereal production is easier than legume production. Most of the cereals were genetically improved so as to satisfy the demands of the population. It is obvious that crop growth is affected by several physiological processes that are sensitive to the wind speed, moisture, solar radiation, soil temperature and air temperature. The climate factors can affect and intervene the diseases, pests and weeds. It is to know that the yields can be projected as the frequency of the monsoon increases. It is important to know that all the crop-related activities are linked with the moisture content at that place. It is obvious that it might be hard to frame any adaptive strategies that can completely nullify the dangerous effects of climate. Some of the adaptation strategies can be associated with crop failure, intercropping, enhanced nutrients, stress tolerance and crop relocation. In the case of gene adjustment, we need to consider the symbionts (Andrews et al., 2018).

Most of the uncertainties can be linked with the enhanced carbon dioxide levels in the atmosphere as it will affect most of the agricultural conditions. Carbon dioxide can be classified as an essential substrate in the photosynthesis process. In some cases, it is obvious that the enhanced carbon dioxide levels can the growth and yields of crops. In the case of legumes, it is evident and we can also observe the nitrogen fixation with the symbionts (Jouko et al., 2007). In the case of water-stressed conditions also, we can observe the carbon dioxide effect. The water use efficiency can be enhanced due to high carbon dioxide levels. The transcription process in the plants will be lowered and this will also reduce the conductance in stomata (Colin et al., 1996). It is still unclear to determine exactly the role of carbon dioxide as there are other abiotic factors in play (Sakar et al., 1988). The cool season legumes are less affected by climate change variabilities.

8.2 MODELING

We know that plant growth kinetics is based on weather factors and diverse physiological conditions that may exist in time and space. The climatic effects on the crop plants can be represented by the numerical models and they may need some calibrations. These models can help us in quantifying the effects of climate variables on various growth stages of plants. These models can help us determine the kind of climatic factors that interact with crop production. These models can help us to predict the crop production rates without much experimentation as it demands more workforce and budget. The International Benchmark Sites Network for Agrotechnology Transfer or IBSNAT was one such model that had wide usage distribution all over (IBSNAT, 1993).

These models of IBSNAT can be classified as mechanistic models and they are based on the equations that can associate climate drivers with the crop growth patterns. CERES of wheat is one such model that comes under IBSNAT and it was validated with extreme confidence (Ritchie et al., 1985). This model can be used to know the effect of climate change on the food supply chains. As described elsewhere in this book, APSIM or Agricultural production systems simulator can be used for a variety of modeling applications. This framework can have several other sub-models that can be fabricated and adjusted according to the needs. With the unique model structure of these models, we can easily simulate the crops (Herman et al., 2013). We can even analyze the farm systems and can suggest the farmers the strategies needed whilst pasture and crop rotations. Food and Agricultural Organization has formulated several models that are based on water and they named them as Aqua-crop (Steduto et al., 2002). Aqua-crop can be based on variables such as canopy cover, thermal time and transpiration. This can be used in several developing countries as a decision support tool. Lentil was taken as an example in this case. Lentils can be called as Lens culinaris. This can be classified as a major cool season grain legume. This can be used for human consumption. LENMOD is one model that can help us to simulate the lentil crop growth. This model can be called as lentil crop growth model.

This model was developed in New Zealand. This model utilized data from the experiments conducted in the summer season, autumn and winter seasons. The sowing dates and other data of many years were collected and injected into this model. It is important to record disease incidence, weeds, soil fertility, potential evapo-transpiration, precipitation, solar radiation and minimum-maximum temperature of the air. Most of the stages in the development that are to be considered in this model can be based on the thermal time that is being accumulated. This model may not consider the stages that involve the emergence of flowering. In this case, the emergence of flowering can be based on the photothermal time that is being accumulated.

The meteorological observations that are to be made daily are maximum temperature, minimum temperature, wet-bulb temperature, dry-bulb temperature, precipitation, duration of sunshine and wind speed. The climate model may need potential evapotranspiration, net radiation, solar radiation and duration of daylight. In this LENMOD model, we can get the seed yield, maturity, leaf expansion, flowering ,etc. The crop growth models can be used to estimate the yields of the crops. This model can be used to estimate the crop growth rate, total dry matter, radiation absorbed or intercepted, maximum leaf area index and flowering dates. The agricultural practices can define the food production levels and it will affect the local environment too (Tilman et al., 2019). The green revolution has changed the agricultural landscape to a greater extent. This has enhanced the usage of fertilizers for a better yield. Though environmental constraints surfaced, we have witnessed the luxuriant cop growths and better yields. With no synthetic fertilizers in the history of agriculture, many countries would have faced famine that has had led to malnutrition and death. The legume crops and fertilizer has doubled the biological nitrogen fixation (Peter et al., 2015). Most of the plant breeders force the genetic makeup of the plants to attain the required yields and results whatsoever.

The cost incurred in terms of environmental and economic assets can be due to nitrogen fertilizer usage to a greater extent. This can be considered as an important factor that is to be considered whilst modeling and restructuring it. Biological nitrogen fixation can be considered as an environmentally friendly gesture towards our environment. This will ensure long-term crop productivity and so the sustainability of the crops can be evident. In the process of producing sustainability, it is essential to look for alternatives that can replace nitrogen fertilizers. Biological nitrogen fixation can be considered as an essential alternative (Frans et al., 2015). Most of the legumes grown can have the ability to fix the atmospheric nitrogen in the soils. This can be linked with their association with Rhizobium bacteria. This relationship can be called a symbiotic relationship. This bacterium can be present in the root nodules. When phytochemicals sense the presence of rhizobium, the plants will start producing nodules so that the bacterium can colonize (Senjuti et al., 2015). Because of this biological nitrogen fixation, the costs incurred in the production of synthetic fertilizers will be reduced drastically.

Most of the legumes can be adapted with the soils that are well-drained. Some of these legumes can grow in volcanic regions. They can easily grow in soils with a pH range between 4.5 to 6.2. These legumes can grow in limestone soils, calcareous soils, alkaline and heavy clays and sands. The adaptation rates of these crops can vary. The perennial legumes can be present in the environments that are deserts and alpines. Most of the legume genera can grow in alkaline soils. Even in acidic soils, they can grow easily. Due to mutual benefits with rhizobia, the mineral deficiency in the legumes can be reduced. In the case of soils with low fertility and dry, we can use green manure to amend the soils. This manure can be produced by legume crops. This can be a solution for long-term problems associated with agriculture. The temperate legumes can be used to close the gap between both summer and winter pastures. The optimum growth of the temperate legumes can occur in the spring time. The advantages of these legumes can be diverse.

They can be salt tolerant. The legume dependent agriculture can be of high prominence as it has more advantages than other forms of agriculture. Hence, several studies were performed to understand the role of legumes. The plant breeders may be directed to work on the constraints like disease resistance, seed quality, maturity and growth habit. It is important for any plant breeder to focus on physiological efficiency. This will be useful in efficient partitioning, higher growth rates and enhanced biomass. The plant breeders should work on some of the constraints like disease resistance, quality of the seed, maturity and growth habit. The selection of legume crops and connected breeding programs can need biochemical traits. Abscisic acid or ABA can be used by the plants in case of stress. The production of ABA can be regulated by the genes and their differences. Excess accumulation of ABS can exhibit greater sensitivity towards water stress. To study drought tolerance, it is hard to define the optimal balance between the factors that regulate water conservation. Water availability can be considered as an important factor in the present food supply scenarios. Water is being used by other sectors for various purposes like industry and urban use resulting in a shortage of water. The water in the sub-soil and surface can be called blue water. This demand can be high in the near future. This will enhance the cost of irrigation and hence the agricultural costs will be increased. Water can be considered as an important factor in the agricultural sector as most of the cellular functions that occur in plants need water. This essential element can be needed in the storage of carbon dioxide. This can be called as carbon dioxide fixation. The stress conditions linked with water can occur due to excess use by other plants or diverted for other purposes. Water deficit stress can be due to decreased water content in the soils. This can lead to drought stress. There are many physiological mechanisms that need water. Most of these correlations were established via research studies. In these studies, most of the physiological characteristics are associated with the mesophyll tissues of the plant cells. With the data obtained post analyzing the physiological activities, hierarchical modes were framed. These models are considered important in the modeling process.

The sensitivities are classified on the basis of the physiological activities. Some of the important sensitivities are to be considered are nitrate reduction, protochlorophyll formation, synthesis of cell wall and expansion of the cells. We need to consider turgor pressure as an important factor that can define the water stress levels in the crops. The turgor pressure is based on the cell membranes and cell walls. The expansion of the cells can be based on the pressure of the cell. It is also based on the yield of the cell walls. There will be cell wall threshold values that can determine the turgor pressures. If there is minimal deviation in the turgor pressure, the permeability of the membranes will be affected. The permeability of the membranes to the water and other ions will be affected. The structure of the cell membrane, receptor proteins, cellulose synthesis rosettes, transport channels and spatial arrangement of enzymes can be regulated by the fluctuations in the turgor pressure. The turgor pressure will get decreased leading to a change in the spatial activities of the protein conformation and so the function of the membranes will be impaired or less than perfect. This will disturb the entire cell structure and its function. There will be deep fluctuations in the cell environment as a response to the local temperature and environments. We can expect many anatomical, morphological and metabolic adaptations according to the climate anomalies of the locations.

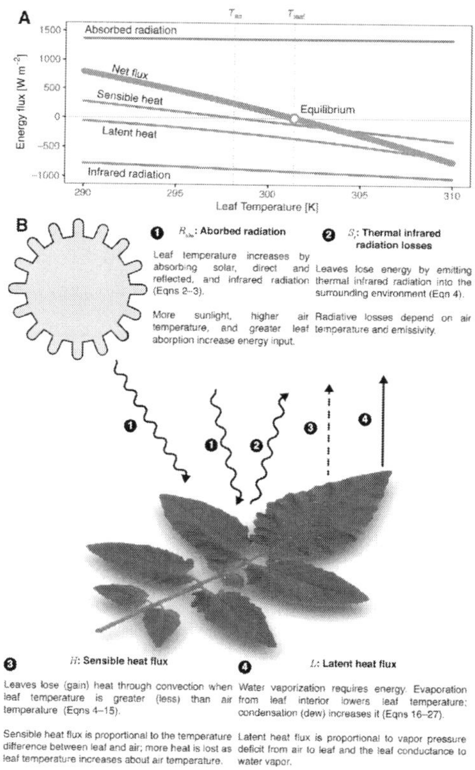

Figure 8.3: Tissue energy balance.

Source: https://www.researchgate.net/publication/338018494_tealeaves_an_R_package_for_modeling_leaf_temperature_using_energy_budgets/figures?lo=1&utm_source=google&utm_medium=organic

Tissue energy balance (Figure 8.3) can be considered as an important factor that deals with the absorption of radiation that falls on the surface of the tissues. The radiation may reach the surface of the leaf from diverse sources. The radiation can be diffuse or direct that may fall on the surface of the leaf. The temperature fluctuations can affect the yield and we need to consider the magnitude of the intensities to determine the crop yields. Less is known about the differential expression in the legume crops. The genetic improvement studies can be used to correlate the sensitivities with the various stages and processes of the plants.

The food legumes can be classified as minor crops though they can be useful for humans as suppliers of a protein-rich diet. In these plants, biological nitrogen fixation can be considered as an important factor. These crops can be grown as secondary crops in both semi-arid and arid zones or regions. These zones will have poor soil moisture and edaphic conditions. They can adapt to the climatic fluctuations by growing non-shattering pods, absence of germination inhibitors and larger seed size (Zohary et al., 1963). The food legumes can be relatively less consumed than the cereals. The pulse crops can be classified as cool season crops and warm season crops. The cool season pulse crops are chickpea, lupins and dry peas lentil. The warm-season pulse crops are cowpea, urad beans, mung bean, pigeon pea, soybean and common bean. These crops can have the ability to grow in frost and cold conditions. Lentils can be considered as having less tolerance to high temperatures. This can be exhibited by the pod set and flowering. The abiotic stresses can affect growth and also initiate the adaptation strategies like bringing up changes in the molecular reactions, biochemical, physiological and morphological fluctuations are among them. The saline soils, temperature extremes and drought conditions can be considered as important abiotic stress that any plant can endure. Most of these crops can be exposed to diverse stress conditions. The plants can sense the climate and weather variations and can exhibit adaptation strategies.

The reduction of the yields can be attributed to several factors that can be complex. These inherent variations can be observed in their developmental patterns and crop architecture. Genetic manipulation can be considered as an important aspect in the crop improvement programs. The agricultural environments can offer the plants complex situations pushing the plants towards adaptation. It is essential to take into consideration the relative differences that might occur in the plant systems. Climate can be considered as an essential factor that can bring about seasonal and regional adaptations. These adaptations may fluctuate the crop yields. The yield rates can be high or low and it is based on the type of adaptation strategy taken by the plants. The improvement in the adaptation strategies can be considered as a basic step in crop improvement. The climatic factors can affect food legumes. This is evident with the water shortage, day length, temperature and insolation. The crops can endure several abiotic stresses whilst ontogeny. This can be evident in the case of temperature stress conditions, soil salinity stress and water stress conditions. The crop yields can be higher in several cropping zones in case of rich water availability. Most of the legumes can be affected by abiotic stresses like mineral toxicity, salinity, waterlogging, chilling, frost, heat and drought (Stoddard et al., 2010).

The ecosystems can be vulnerable to flooding events. These lands can be called as wetlands. These wetlands can support a variety of species in the ecosystem. Several plant varieties can be present in these wetlands. The prolonged presence of water can lead to biotic and abiotic stresses in plants with a main focus on the rhizosphere region. The soil microbes can enhance their population and so the soil chemistry will be altered. This will demand altered metabolism in the plants. Even the soil flora will be altered. Due to these conditions, the performance of the plant and root growth can be affected. Waterlogging conditions can result if soils withhold water for prolonged periods and this might occur due to diminished drainage. We can observe that the land use change has affected the wetland crops. The irrigation systems on a large scale and intensive irrigation of the farmlands can enhance the waterlogging conditions in the soils. There are many factors that determine flood ecosystems. Due to prolonged flood ecosystems, there will be lowered oxygen levels or partial pressures at the root zones. This is considered as important due to flood-related plant problems. Most of the roots can be sensitive to the anaerobic or anoxygenic conditions in the soils. These anoxygenic conditions can affect the microbial communities and their growth rates. Due to this, most of the soil nutrient dynamics will be affected (Nguyen et al., 2014). Due to the incidence of anaerobic conditions in the soil, most of the plants suffer from impaired aerobic respiration.

The aerobic respiration in the plant roots can be efficient but in case of flooded conditions, the root cells will switch to the anaerobic conditions. This anaerobic respiration can be based on the pH levels. When the pH values are greater than the neutral range or 7, we can observe lactate fermentation. This fermentation can be dominant in anaerobic conditions. If pH values fall below the threshold levels, there will be an onset of ethanol fermentation. If the pH falls well below the pH of the cytosol, then we call it as acidosis. This is the main problem associated with plants if they are exposed to flooded conditions. In the case of plants with flood tolerance, the drop in the pH values will be neutralized by the alkalization mechanisms. In the prolonged anaerobic conditions, the roots will start accumulating arginine, amides and alpha-aminobutyric acid. It was observed that in hypoxia conditions, there was an accumulation of lactate, ethanol and pyruvate. Prolonged flooding can enhance the activity of alcohol dehydrogenase and this is evident in the case of clover roots. The formation of lactate dehydrogenase can be considered as an important sign in determining the injury caused to seeds by floods. The peroxides and catalase activity can be evident in the coleoptiles of rice crops. There will be chlorosis in the lower leaves of the sunflower plants that are living in the flooded conditions. This situation can be due to reduced cytokinin import from the roots that were flooded. The production of polysaccharides can be lowered due to the altered activity of the P metabolism. The reduced oxygen levels and low temperature of the soils can reduce the development of shoots. This proves that the development of shoots will be controlled by the roots.

Due to increased water content, the production yields of the food legumes will be reduced. There will be reduced crop growth, chlorosis and reduced nitrogen fixation. Most of the legumes are tolerant to waterlogging conditions if it is temporary. Some of the legumes that are tolerant to water logging conditions are moth bean, grass pea, pigeon pea and mung bean. Pigeon pea can be classified as a sensitive species that suffers with waterlogging conditions. This is the reason why this species cannot be grown in all areas. Ethylene production can be due to the microflora of the soil and this can cause considerable damage to the crops. The intermittent waterlogging can affect the biological nitrogen fixation mechanisms and there will be severe N deficiency. There are several genetic differences that can be observed whilst studying physiological tolerance. The water availability for agriculture can be in demand always like most of the water is being used by other sectors. Most of the water relations can be linked with physiological processes and cell growth. This will determine the water potential. We may not get the exact value pertaining to the water potential. The turgor pressure present in the corn tissues can vary with the developmental phases.

Some of the physiological processes that are associated with the water potential are cell growth, synthesis of the cell wall, protein synthesis, formation of protochlorophyll, nitrate reductase, stomatal closure, carbon assimilation, sugar accumulation and proline accumulation. Turgor pressure can be classified as an essential indicator to represent the water stress conditions of a plant. Reduced water potential in the mesophyll cells can affect the cell physiology. This lowered water potential can lower the chemical activity and so the cell structure. The efficacy of the cells will be reduced as the hydration of proteins will be impaired. As the position of the cells changes with water shortage or deficiency, the relationship between plasmalemma, tonoplast, endoplasmic reticulum, mitochondria, nucleus and chloroplast will change. If there is a loss in the turgor pressure, most of the spatial positions of all the membrane enzymes and transport channels will change and this will reduce the thickness of the membranes. Any variation in the pressure of the cell can render the cell towards shrinkage and so the plasmodesmata will be damaged. The molecular concentration in the subcellular location of the cells can get changes or even damaged in many cases.

The enhanced tissue water deficit can be due to the lowest water supply that can result in the lowering of the turgor pressures. If the cells are expanding, then the turgor potential can be constant and this will lower the water potential. Hence the expansion of the cells will be lowered. Most of the effects of turgor pressure will start from the older cells and reach the developing cells. The primary effect of the water shortage can be due to the closure of stomata. Most of the stomata can get closed due to root signals (Davies et al., 2010). The stomatal cells can get closed if the vapor pressures are high. This may not be linked with the water potential. If water limitation tends to prolong, photosynthesis can get decreased along with respiration. So, it is obvious that the ratio between these two processes will get decreased. There will be an increase in dark respiration and photo-respiration and this will push the plants to starve. The shoot systems can be damaged and so their metabolic effects. These metabolic defects caused due to water limitation may be devoid of the factors that affect carbohydrate deprivation. The translocation of the carbohydrates whilst water shortage can be considered as an important factor. The phloem loading may be resistant to the water deficit conditions (Sung et al., 2018). The translocation of the photosynthate translocation can be attributed to the variations in the source-sink interactions. These interactions between source and sink can be due to water stress conditions. There will be a lowered flow of carbohydrates in the phloem tissues. It was observed that minimal correlation exists between the leaf area and photosynthetic rates. The drought conditions can be linked with vigor (Leport et al., 2006).

8.3 CLIMATE CHANGE IMPACTS

The ill effects of climate change can be immense on humans. This will lead to diseases in humans. Due to global warming, the intensity of disease at one place may spread to another place easily. Due to enhanced carbon dioxide levels, most of the plant populations will become sensitive to diseases. There are additional effects of global warming. The organisms that are sensitive to diseases can become more vulnerable. Due to climate change, there will be variable growth rates. The higher temperatures can damage the pathogens. In some cases, pathogens like insects may escape and will reach other geographical areas. In general, the high temperatures can damage fungal pathogens, especially in the secondary hosts. It is obvious that the local conditions will be defined by both environment and geography.

Water scarcity (Figure 8.4) can be classified as an important limiting factor in the production of crops. This problem is dominant in the semi-arid and arid regions. Drought can affect most of the processes of metabolism and growth of the plants. Drought stress can be based on the exposure time, intensity and crop growth. If moisture is less, there will be no use of using high yielding varieties and fertilizer application. The availability of nutrients can be classified as an important factor in the growth and productivity of the plants.

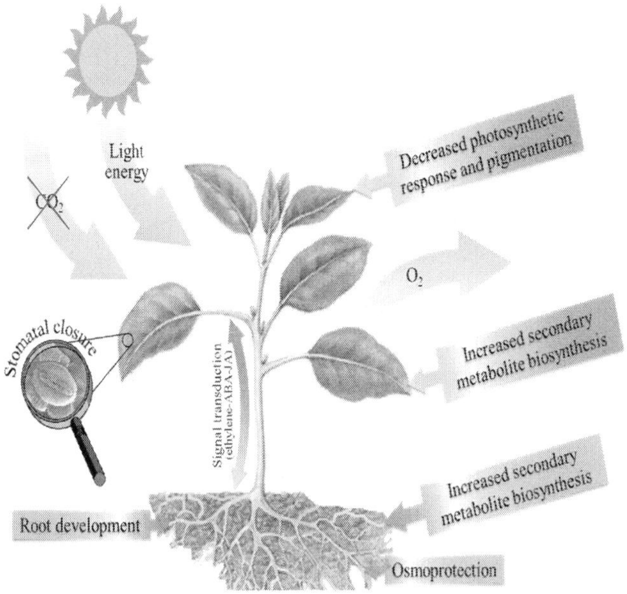

Figure 8.4: Water scarcity in plants.

Source: https://www.technologytimes.pk/2014/12/09/impact-of-heat-amp-drought-stress-on-physiological-biological-processes-in-plants/

The nutrient uptake of plants can be classified as an important process. The plants can exhibit a root system that can be effective. This can be in the form of root length, density of root hairs, radius of roots and root extension. Legumes can be extremely sensitive to moisture stress and it can be explained in terms of both growth and productivity. Half of the legume producing countries face soil moisture stress. Due to prolonged drought conditions, the nodule formation can be inhibited (Streeter, 1995). Due to this, the bacterial colonies may not grow and hence the symbiotic association will be reduced leading to anomalous nitrogen fixation. There are several nutrient deficiencies that can be observed in legume crops. The legume crops may not be useful in improving the physical conditions of the soil but the nitrogen levels in the soil. The legume crops may need an excess amount of P for both growth and productivity. These crops can be less tolerant to P stress conditions. (Figure 8.5)

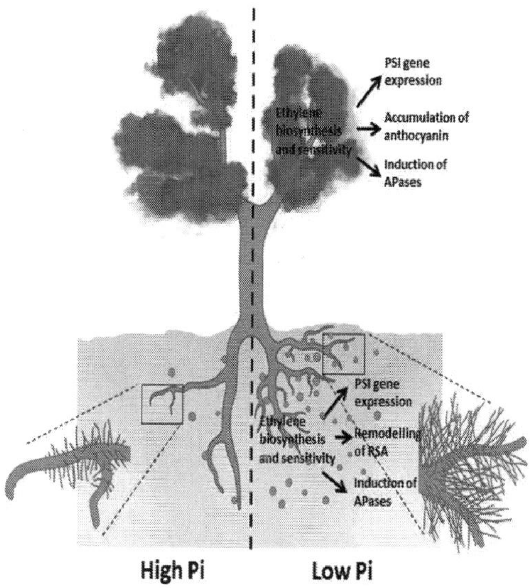

Figure 8.5: Phosphorus stress.

Source: https://www.frontiersin.org/articles/10.3389/fpls.2015.00796/full

The application of P can be useful in enhancing the productivity of the legumes. This is true even if the legumes are in water-stressed conditions. In the case of lablab crop, the enhanced levels of P increased dry matter. The plants can absorb P as phosphate ions and there is a possibility that less amount of organic P can be absorbed by this plant. P can undergo many changes and so it might not be available to the plants. There are several physiochemical changes in P molecules that need optimal pH. In the case of low pH conditions, iron and aluminum suppress the P mobilization. At alkaline and neutral pH, most of the calcium compounds can associate with P and so the availability of it will be drastically reduced. The presence of organic matter can be important in the P absorption. Some of the organic compounds can encourage P absorption. Ions compete for some site of P. Most of the P fertilizers can be soluble and we can easily use it in the form of fertilizers. The P fertilizers can be used before the sowing time so that they can be easily absorbed by the roots. The P levels can be maintained properly as it is classified as an essential ingredient that can generate ATP which is an energy currency in plants and other organisms. P can play an important role in both the growth and survival of the plant seedlings. Plants can acidify the rhizosphere and it will enhance the solubility of P. Some of the legumes like faba beans, alfalfa and clover can generate acidity whilst N fixation in the drought stress conditions. When the N fertilizers are added to the soils, there will be acidification. If the soil moisture is less, then we need to add N and P fertilizers so that the yields will be increased. The drought stress tolerance can be seen in plants that were fertilized with P. The drought stress can enhance the N concentration in proteins and shoot in the grain crops.

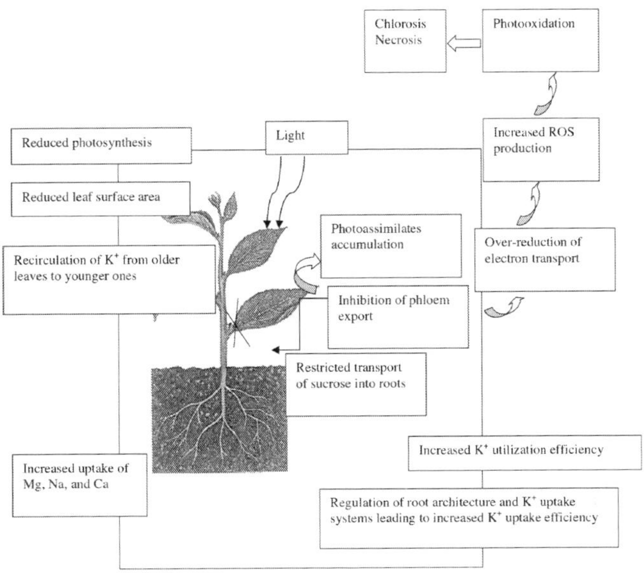

Figure 8.6: Potassium in plants.

Source: https://link.springer.com/article/10.1007/s11738-014-1491-2

Potassium (Figure 8.6) has diverse effects on plant metabolism. This element will enhance resistance towards abiotic stresses. Decreased uptake of K can lower the photosynthesis rates. The dark respiration rate will be increased and so are the assimilates. K application can enhance plant productivity and photosynthesis rates. K fertilizer can lower the effects caused due to water stress conditions. Even in saline conditions, K plays a crucial role in growth and development. K can be important in controlling osmoregulation. This will affect stomatal closure.

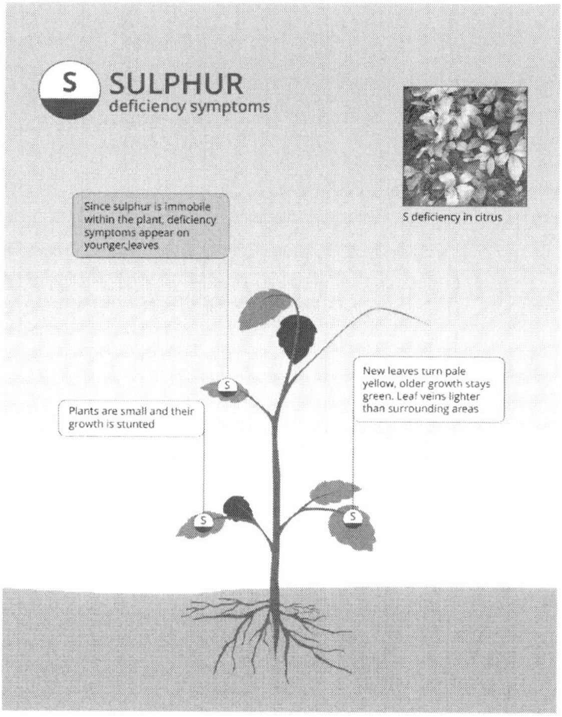

Figure 8.7: Sulfur in plants.

Source: https://www.iclfertilizers.com/sulphur

Sulfur (Figure 8.7) can be classified as an essential element that can participate in several metabolic reactions. The production of proteins and fatty acids will be on the supply of S molecules. The S element can be important in the production of the sulfur-containing amino acids like thiamine, methionine and cystine. Most of the arid regions have S deficiency. This can be due to a reduced supply of fertilizers and organic matter. High cropping intensity, intensive cultivation and chemical fertilizers can result in

S deficiency. High S rates can be toxic and that can reduce crop productivity. Calcium and magnesium can be important in maintaining membrane permeability and cell integrity. These two elements can enhance the growth and germination of the plants. The cell elongation, division and mitosis will be enhanced leading to accelerated plant growth. These elements can participate in protein synthesis. These elements can be useful in the transfer of the carbohydrates that might be produced whilst photosynthesis. Calcium can detoxify the effects caused due to heavy metals. It is important that calcium is maintained in appropriate concentrations in the plants. Due to this, the productivity and growth rate of the plants will be increased even in the water stress conditions.

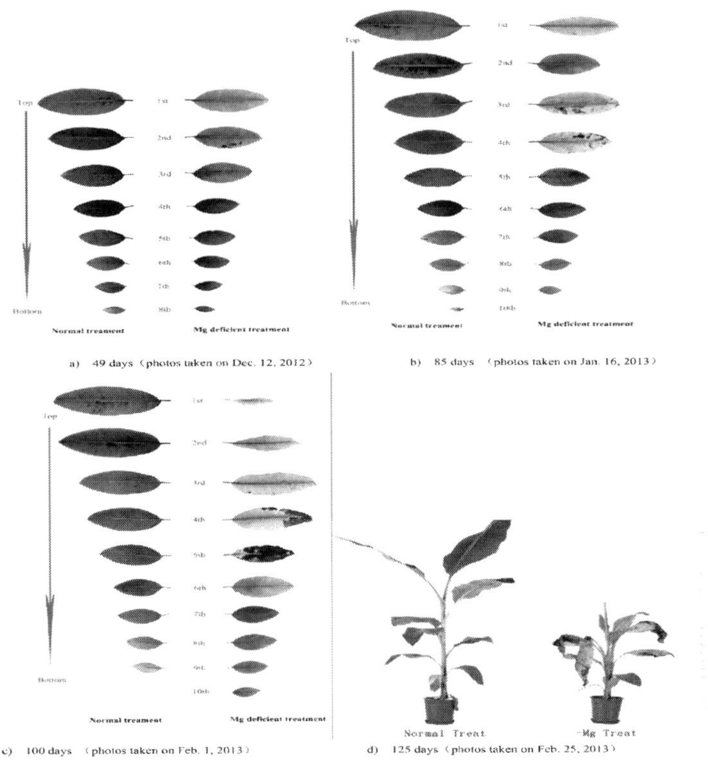

Figure 8.8: Magnesium in plants.

Source: https://www.iclfertilizers.com/sulphur

Magnesium (Figure 8.8) can be considered as a mobile element and its deficiency may not be neutralized easily. The main symptoms of this deficiency are necrotic spots, yellowing and chlorosis of leaves. These spots can be normal in case of water deficit conditions. The sodicity and soil salinity can be considered as important problems in the agriculture sector. The soils that are affected by the salts can be classified into naturally occurring salt enriched soils and manmade salt enriched soils. The soils that have excess salts can damage plant growth. In this case, the organic matter will turn black or dark brown. The soil structure will get deteriorated and so the physical characteristics. It is difficult to rejuvenate these soils. The water salinity and soil salinity stress can have a major effect on plant growth. The nutrient ion distribution will be disturbed and there will be an ionic imbalance. The suppression of plant growth can be seen more frequently. The lethal salt concentrations can affect the plants in several ways. Infertility in the arid zones can be attributed to the high salt concentrations. There will be nutritional imbalances. This can be considered as a direct effect of sodicity and salinity. Roots can affect first if there is a high salt concentration. The salts can hinder the development of the root radical whilst the germination process. The legume roots can be stopped to grow and the root hairs might get damaged. We can observe root mortality in acute salt concentrations. In the case of soil hardening, the nutrient uptake of ions and other essential ions will be hindered. Root respiration will be impaired. Sodication, salinization and salt accumulation can be important factors that can affect the growth of legumes. It is essential that the salts that are accumulated should get percolated into the interior layers of the soils. In some cases, certain crops can grow even in the salt-stressed conditions and this is due to the fact that most of the salinity will be taken away by the rainfall.

Chapter 9

Geosensors in Crop Variability Studies

CONTENTS

9.1 INTRODUCTION

The sensor technology can be advantageous if it has high temporal (Figure 9.2) and spatial resolution (Figure 9.1) and if it has data transfer capabilities in real-time. This will ensure that the high-resolution data sets reach the end-users in the stipulated time. This will avoid unnecessary costs incurred whilst framing policies or changing them according to the current weather scenarios. It is obvious that agricultural production became a fabric of our human ecosystem for ages.

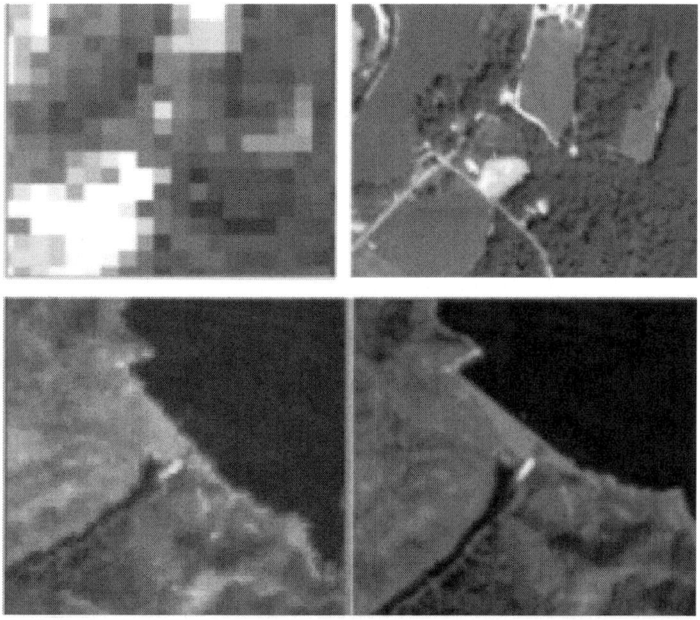

Figure 9.1: Spatial resolution.

Source: https://www.slideshare.net/NabeelKHALID13/spatial-resolution

We relied on it for several causes and yet in years to come we follow the same trend. Most of the crop production can exhibit variation in both space and time and this can be attributed to the weather conditions. This can be based on the external factors and management practices being followed at present times. The information perceived from the sensors can be important so that the farmers can be warned well before the devastating event happen. In the early ages of agriculture, humans relied on direct observations and made records so that in case of recurrence, we can act accordingly. In the last decade, several advancements arose and it led to the production of airborne sensors, satellites, weather stations and soil moisture sensors. All these advanced systems and related products are being adopted by many farmers worldwide.

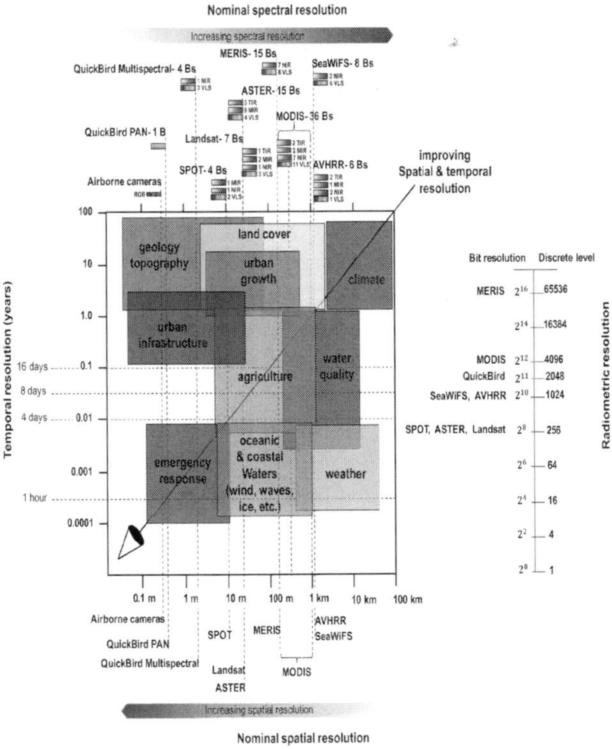

Figure 9.2: Temporal resolution.

Source: https://www.researchgate.net/publication/308514016_Advances_in_remote_sensing_applications_for_urban_sustainability/figures?lo=1

The sensor networks and automatic sensor systems help the local agricultural communities in several ways. The near-time observation capabilities helped most of the researchers in forecasting the problems associated with agricultural practices. This had led to the production of sustainable methodologies thereby lowering the negative impacts on the environment because of agriculture. Most of the risks associated with food safety were lowered. The communication and sensor technology emerged to develop offline sensors and this was assisted with the downloading and field logging capabilities. These resources led to the generation of the autonomous sensor webs and they are interoperable in nature. The concept of sensor web can be based on SWE or Sensor Web Enablement framework. This framework is a part of the OGC or Open Geospatial Consortium. (Figure 9.3)

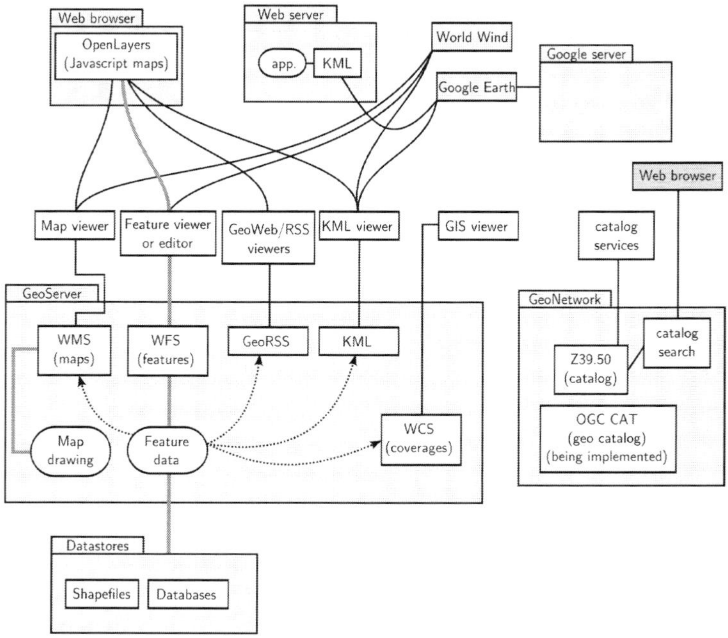

Figure 9.3: Open Geospatial Consortium.

Source: https://en.wikipedia.org/wiki/Open_Geospatial_Consortium#/media/File:GeoServer_GeoNetwork_with_web_app.svg

There are several protocols, web services and interfaces that can be used to exchange data and process it. The data available can be shared across different end users. There are many sensor networks and sensors associated with the development of the data sets that can be hoisted on web services. The geo-sensors can be defined as sensors that are capable of monitoring any phenomenon in a specific geographical place (Nittel et al., 2009). Most of the attempts made to develop sensors and their associated networks were associated with technological up gradation and innovations. This technology rush has helped many farmers to grow the food and fiber so that the population demands will be met. The agricultural field can get benefitted from the high-resolution data sets as this can enable real-time data transfer and help us to know and quantify the in-field variations.

There are several farm decisions that can be based on the weather patterns either completely or partly. This is due to the fact that weather is important for every crop. The information pertaining to weather can be considered as an important element in almost every process involved in modeling. This is the reason why the agro-meteorological networks are placed. These networks can provide weather data in an integrated manner. This data includes agricultural data along with soil data and weather data. With this data, we can produce crop models. The pest information and plant disease information can be delivered by these services so as to provide a base for several researchers and planners to get their model tested or policies verified in time prior to their approval. We can use these models in the irrigation processes and plant protection systems. There are many countries that have the infrastructure to deliver agro-meteorological data to the farmers beforehand. Some of the data providing stations might need an upgrade as the latest technologies provide accurate data than conventional technologies. Most of the data obtained from these sources will be injected into the agricultural models and so we can generate several management decisions that can help the farmers to get their yields in time and without hardships whatsoever incurred whilst farming process.

In the U.S, there are several agro-meteorological stations that can provide us data needed for the forecast and prediction for at least a week or more. In this case, biological knowledge-based will be used along with the data obtained from the sensors. This can be used in the decision-making process. There are several applications that rely on the data obtained from the sensors. Some of them are manure supply and spread, monitoring of crop load and performance of the irrigation systems. It is essential to know the soil attributes along with the spatial variability in the fields. This can be important in the management of site-specific problems. Prior to the sowing period, the farmers tend to select appropriate crop types and the soil conditions so as to deliberate their choices for getting a better yield. The sensor networks can provide him/her the data and the reports pertaining to his/her needs. This will increase the comfort levels of the farmer and he will be having an estimate about the problems associated with the crop type chosen. Whilst the growing period, the farmers might demand information about the pesticides and nutrients. The timing of their application can be important for any farmer as their application at the wrong time can lead to severe distress.

There are many methods to sense the physical properties and chemical properties of the soils. We can broadly classify these sensing techniques into proximal sensors and wireless sensor networks. The proximal sensors can be installed on the vehicles and we can get information according to an 'on-the- go' basis. The wireless sensor networks can be a network of individual nodes that may be located at various positions and depths in the parcel in which we are interested. Most of the proximal sensing techniques can be classified as field-based techniques and these techniques can be used in the quantitative analysis of mineralogical, physical, chemical and biological properties from a distance set by the user. In general, 2 meters can be defined as an appropriate level at which we can get the data with more precision. Most of these proximity sensing techniques are based on diffuse reflectance spectroscopy and visible-NIR spectroscopy. We can even use Mid-IR wavelengths for that matter. In some cases gamma-ray radiometry can be used. We can use ground-penetrating radar or GPR along with electrical conductivity sensors to know the parameters of the soils. We can get information about the surface soil and other properties also. We can get the pH, nitrogen content and organic carbon of the soils. The accuracy will be based on the soil property that we are interested in. EC and GRR can give us information about the subsurface soil layers. We can get textural information about the soils. This can be classified as an important measure in the sensing process as far as crop sensing is concerned. Ground penetration radar can be used to quantify the subsurface strata of the soils. It also gives us information about soil moisture levels.

We can integrate several sensing techniques and bring about an accurate dataset that can be used by the research communities and farmers. We can use multi-sensor platforms to get accurate information about the soils and their properties. These platforms can help us to get the values with extremes accuracy and precision. This is due to the fact that independent measurements will be made with variable spectral regions. Each value will be unique and hence error rate will come down drastically. Soil compaction problems can be easily monitored with these sensing techniques. The soil compaction problem can be considered as an important factor in crop production. Wireless sensor networks or WSN can be used to know the soil properties. We can get the soil temperature and soil water content with these networks. This is most reliable because we get real-time data from the in-situ sensors. This will reflect the accurate field conditions so the modeling process will become much easier than ever. The surface-to-depth profiles can be easily obtained with these techniques and can be supplied to the end users easily. These measurements can be easily used in the form of input in the farm management practices. We can use the wireless sensor web to get the measurements in the optimal wavelengths. This will also provide us surface-to depth properties of the soils. The scheduling and estimation parameters of the sensors should be set prior to the analyses. The communication protocol and data transfer algorithms must be taken into consideration at all times necessary. We need to deploy a wireless cell with solar power in order to get the data at all times. The data-logger, communication modules and optical sensors should be used. Most of the researchers suggest that the cells powered with solar panels can be used in the agricultural fields. The proximal sensing systems can be used in various operations as there will be real-time data transfer. This data can be used in various phases of agriculture. The techniques that are being used in the analysis need to be advanced and accurate. These sensing systems can enhance the efficiency of the data collection process. This data will provide us information about the spatial patterns pertaining to the soil variations. The spatial management of the stratification zones and crop yield patterns can be studied using data obtained by these techniques. Latest advancements in imaging technologies have helped us to know the soil properties from a distance decided by the client or user. The electrical resistance sensors, dielectric sensors, tensiometers can be used to know the soil properties remotely. It is essential to take due consideration whilst installing the sensors. The control of weeds, insect pests and diseases can be considered as an important operation that can help us to get better yields. It is important to have actuators and sensors that can be attached to the tractors or any farm machinery to get accurate results.

We can use spectral color analysis and associated techniques to know the textures of leaves. This technology was used in the analysis of weeds and their differentiation from healthy crops. Post sensing the disease and weeds, it is important to use thermal methods, chemical and mechanical methods to control both weeds and disease. Several advances were made in the production of the proximal remote systems. Some of them can be installed on the farm machinery and even hand-held systems are in place. If the irrigation system should be perfect and worthy, then we need to use the proximal sensing equipment to ensure any problems. RFLID technology can be used to send the data easily over a network at great seeds. This will reduce the costs and so the power usage will be less. Variable rate irrigation or VRI can be classified as an important system to know the exact amounts of water utility. This technique or system was first used in the U.S. The irrigation systems can be regulated by the sensor data that was obtained from the real-time sensors. The irrigation systems can be automated and this will allow diverse inputs. This system can analyze the nozzles and the location-based problems can be sorted out easily. The decision support software can be used in the regulation of irrigation and it will be based on real-time sensors. Irrigation management systems demand communication and sensor technology. The sensors should be easy to install and the cost of the instruments should be minimal. The interoperability of the sensors should be considered always and this will decide the efficacy of the sensors.

Remote sensing can be used to know the water status present in the crops. We can use thermal imagery based on spatial information. The mapping and yield prediction of the crops can be monitored by remote sensing devices. We can use these monitoring systems to monitor the yields. The crop quality can be used to analyze the crop yields. Some of the remote sensing applications can get us information about the protein content and crop moisture. The qualitative analysis can also help us to know the crop variability. This will help us understand the nitrogen fertilizer requirements. This will be useful in the environmental auditing process. This process can be done on a field scale or as per the user requirements. In precision farming, the sensing of crop quality can be considered as an important aspect. The fleet management and control systems or FMS can be used in the management of mobile vehicles and other remote sensing devices. This FMS can have immense uses in the agriculture sector. This will enhance the production of the crops. In this case, we have two important components. They are telematics and software needed to analyze the data obtained from the remote sensing systems. This software can be used with the wireless networks at a region specified by the users. The FMS system can be linked with the internet to share the data as and when needed. The intensive crop production systems may need heavy machinery. It is important that they need to be maintained at all times.

9.2 YIELD VARIABILITY

Remote sensing tools can be used in the analysis of crop yields. These tools can reduce the gaps between the data and can enhance our knowledge on the problems associated with crops like diseases and weeds. Researchers have made many efforts to get the required information on the satellites that can be used in improving crop yields. The main parameter that can be studied with this data obtained from the satellite image is yield. This can be defined as the grains produced per land unit area. We can use the related images obtained from the satellites for this purpose. The satellite data can be used to know the yields with high accuracy and this can be used for every crop and for every year. The yield related data based on the satellite images can be a challenging task. Some limitations that can be considered whilst determining the accuracy of the image are resolutions like spatial, temporal and spectral resolution that can be provided by the satellites. These relationships between resolutions can be complicated and it needs expert intervention to derive useful information. The main ingredients of this process can be crop productivity and parameters that can be quantified. The yield estimates derived from satellite data can be used for many applications. There is a demand for both spatial and temporal accuracy. The important advancement in this field can be the famine early warning system. This can provide us satellite data in real-time. The remote sensing techniques can be useful in the estimation of the crop yields and this will save most of the costs incurred in field observations. These tools can help us estimate the crop yields at both regional and national scales. These tools can be used in the gap analysis. There are two important characteristics that are to be considered as important. These systems can be operated in locations that cannot be accessible to us in normal conditions. We can get high accuracy in the fields. We can reduce both time and expenditures. In the case of maize crops, we can use reflectance indices to get the crop yields (Walter et al., 1999). These statistical approaches can be useful in analyzing the data obtained from the fields. There are several approaches that are being developed to acquire and analyze the data obtained from remote sensing devices. Light-use-efficiency approach can be considered as a primary approach. With this approach, we can relate the biomass quantity with APAR or absorbed photosynthetically active radiation. The biomass quantification can be done all year. The biomass production can be associated with the constant harvest index or HI. The crop yield can be evaluated by using the product values of APAR, RUE or radiation use efficiency and HI. The RUE values will be high in the case of C_4 plants. In maize crops, this value will be high. This approach can be classified as the APAR method and it is being used by various sensors over different regions with diverse crops. We can use the APAR approach to get observations frequently. This is true if we want to get data for 8 days and 16 days composites. This type of data can be obtained from MODIS. We can use the gross primary production product of MODIS to know the plant growth rates.

We can use crop simulation models also. These models can be used to predict and analyze crop yields. We can manipulate the data based on the parameters obtained on a pixel-by-pixel basis. This approach can be employed in the regression steps. We can know about multiple predictors. This can be done only if the images are available for more dates for a specific region. The accuracy of the data generated from the crop models can be used in the simulation process. Some models can simulate biological stress accurately. In this case, we can consider weeds and pests as important biotic factors. These models cannot take into consideration salinity or any other abiotic factors. We can use two types of satellite data for this purpose. We can get Landsat and MODIS data. MODIS data can give us 16 day MVC or maximum value composite products. The resolution of this data is about 250m. This can be applicable to both Aqua and Terra platforms. WE can use enhanced vegetation index or NDVI for this purpose instead of choosing the former type of data. In this case, we can use red and NIR spectral bands. These spectral reflectance values can be used to compute the vegetation indices. This can be called as the wide dynamic range vegetation index or WDRVI. This index can be used in the case of maize crop canopy. The MVC products can be used with coarse filters. This will drastically reduce the error propagation and it is true with the vegetation greenness. This will lead to negative biases. If we want to create time -study, it is essential to use actual composite date data than the composite center window. This will be accurate and give us the vegetation index with minimal errors. IN order to reduce the error propagation, we can use MODIS data with a daily time scale. For this, we can use local polynomial fit (Zeng et al., 2009).

The MODIS sensors can be classified into Aqua and Terra. The spatial resolution will be 250m. The temporal frequency can be 16 days and we can have an 8-day offset between these two platform sensors i.e. Terra and Aqua. The variables that are to be considered are the NIR band and red band reflectance values. The red band reflectance can be between 620 to 670nm. The near- infrared or NIR can be between 841 to 876 nm. Landsat sensors can be used in this case. Landsat sensors that can be used for this purpose are Thematic Mapper or TM and Enhanced Thematic Mapper or ETM. The spatial resolution will be 30m. We can use the red band and NIR band. The red band reflectance can be between 630 to 690nm and NIR band reflectance can be from 760 to 900nm. NASA POWER is an important source of data. This can provide us agroclimatology data. We can get daily data at a spatial resolution of 2^0. The important variables that can be derived from this source are wind speed, relative humidity and incident solar radiation. In order to get the accurate MODIS pixels based on the field points, we can use geocoded polygons. The polygons can be reprojected to MODIS sinusoidal projection. In general, we can project from WGS84 geographic to sinusoidal projections. It is important to avoid certain pixels that get polluted with the adjacent areas. It can be essential to use pixels that have 80% of the region.

The Landsat data can be important in the analysis of crop yields. The Landsat scenes can be downloaded free of cost and for many areas, we can get cloud-free data. We can download these scenes from USGS EROS archives. These images should be converted or transformed into reflectance values. This can be done using LEDAPA or Landsat Ecosystem Disturbance Adaptive Processing System preprocessing code (Masek et al., 2013). This algorithm can calibrate the scene and can convert TOA or top of atmospheres reflectance. This can be applied whilst atmospheric correction. We can use radiative transfer methodology also (Vermote et al., 1992). We can calculate the surface reflectance values and also can attain mean surface reflectance. This can be done with both Red and NIR bands. We can perform zonal statistics also. This can be done using field boundaries. This can be converted to WRDVI and then we can do analysis. We can use two image dates. We can use scan-line corrector failure software for this purpose. This will remove or neutralize the scan line errors. With this, it is obvious that the mean values can be classified as good pixel values.

9.3 PRECISION AGRICULTURE

Precision agriculture can be important in order to enhance production yields. It is also called as PA. In this technique, we can detect and manage spatial variability. This can be studied as an intra-field spatial variability. This will enhance profitability and lowers the impact on the environment. It can be classified as site-specific crop management. It can be classified as a form of precision agriculture. In this technique, we can classify the fields into zones that can be manageable. This can be used in the analyses of the yield limiting factors. This will be dependent on the quantitative measurement of these factors. The variable seasonal rainfall can have a profound effect on plant available water holding capacity of PAWC. The PA or precision agriculture can recognize and then can help us manage the intra-field variabilities. The data obtained via PA can reduce the load on the environment and hence can lower the environmental problems drastically. It is essential to know SSCM or site-specific crop management and it was classified as a form of PA. This will focus on managing the crop yield variability with the aid of agricultural inputs. This can classify the crop fields according to the user's needs. The spatial variability and the yield variability can be classified as complication interaction. This can be a result of several factors that can affect the growth of crops and those are climatic factors, topographical factors and soil factors. The climatic factors include temperature and rainfall. The topographical factors include elevation and other factors. Some of the soil factors that need to be considered as soil water and nutrients.

The climate , in general, has spatial variability and this can bring both spatial and temporal variability in the crop yields. These yields can differ from season to season. It is essential to know the temporal variabilities that encourage PA over conventional farm management practices. It can be known that the degree of variability and PA can be related or not. There are several issues that are to be considered and fed into the models. So, the SSCM may demand a complete understanding of both spatial and temporal components of crop growth patterns. In some countries, the yield variabilities that are observed can be dependent on soil water availability. The plant available water holding capacity of PAWC can guide us to know the intra-field spatial variability if any. It should be noted that there should be a delineation of zones. We can infer that PAWC can be regulated by the seasonal rainfall. We can say that PAWC associates with rainfall and it can regulate the spatial and temporal components that define the yield variabilities. The delineation of the agricultural zones can be dependent on limiting factors and soils. The yield data can be classified as the first information source. It is important to observe the variable patterns temporally and it should be extended for multi-year maps of crop yield variability. Several researchers developed diverse approaches to observe and analyze yield data from multiple years. We can use temporal variograms and spatial variograms. Here semi-variance can be classified as a function that can define both spatial lag and temporal lag. With these inputs, we can know the spatial trends and temporal maps. The availability of climate data for all the years under consideration can be another limiting factor. So, it can be assumed that remote sensing technology can fill these gaps with the data generated from the images extracted from the archives.

The vegetation indices can be developed based on the data obtained from the remote sensing devices. These indices can evaluate the growth variabilities. This will help us understand the health parameters of the crops. We can also quantify the relative growth patterns. The most popular in this category is NDVI. This index can be used to estimate the vegetative vigor. This can be obtained by evaluating the spectral reflectance values. NDVI can be used in many monitoring applications. We can use single images to estimate crop yields. We can know the growth variability annually. This can be based on the interactions between the soil and climatic conditions at a local scale.

Nitrogen management has been studied with many crops and they are associated with the crop sensors for various applications. The biofuel crops can be used to generate low carbon emissions. The association of crop sensors helped many farmers to get the estimates of the actual nitrogen levels and requirement of nitrogen management practices. Most of these sensors will work in on-the-go mode. The determination of the N rates can be important in the formation of yields as expected. The rate of fertilizer application can be an important tool. This can be manipulated so as to get the required yields. The nitrogen content that is being supplied by the soils derived from the minerals may not suffice the crop requirements. It is essential to use fertilizers to get the required nitrogen levels. The site-specific crop management practices can be framed to optimize the production of crops. This can be done by taking into consideration the soils, crops for each crop field. We can use sensor-based and map-based approaches. These approaches can help us to consider variable rates of fertilizer application. In the case of the mapping approach, we can use a variable rate of fertilizer application, map production and grid soil-plant sampling. The spatial resolution of the area can be considered as an important factor. The sensor-based methods can be based on the reflectance values and with these estimates, we can know the nitrogen status of the plants present in a given area. The data obtained by the sensors will be in real-time. The real-time data can be used to regulate the field operations. We can get the need for N content in real-time. With these advantages, we can estimate the fertilizer needs of the plants and crop fields at high resolution. It is important to use the algorithms needed for the sensors so that they can work according to the user's needs.

The nitrogen management can be variable and its maintenance can be dependent on the reflectance sensor. This can be important in precision agriculture. The data obtained from the reflectance sensor can be used to optimize the efficiency of nitrogen use. With this, we can drastically reduce the dangerous impacts on the environment and improve agriculture. The relationship between agriculture and the sensors can be linked with the growth stages of the crop. There will be minimal saturation when the nitrogen uptake by the plants is higher. The reading generated from the sensors can be from the respective plots. This data can be associated with the crop parameters and the calibration functions. This data can help us to know the nitrogen capacity of the soils. These measurements at the sensor can be used to estimate the crop biomass and the uptake of N by the plants. The improvements in precision agriculture have led to the production of active sensors that are mostly ground-based. These sensors can be used to determine the NDVI values. NDVI can be linked with the nitrogen content of the leaves. This can be observed in the maize crops. The NDVI values that were generated through remote sensing devices can give us more information about the nitrogen variability in the soils. It was observed that there are several relationships between the NDVI values and the total grain yield in maize. There are many studies that were conducted to link active sensors with field conditions. The NDVI can help us understand crop health also. We can determine and compare the efficiency of the two sensors that can be active. The grain yield can be estimated in advance. We can easily understand the crop yield in semi-arid environments also. Earlier, the data pertaining to these indices was procured by passive remote sensing devices.

In the case of advanced sensors, the need to light energy is mandatory. This will help us in extracting the NDVI values at certain locations at specific times. These sensors may not need ambient light for data acquisition and retrieval. The sensors for crop canopy can be small and they have a light source integrated with the machinery. These sensors can be used in the presence of visible light and near-infrared light. Mostly we can use light at 400 to 700nm and from 700 to 1300nm. This can be used in the estimation of properties pertaining to crop canopy. The reflectance of visible light can be based on the chlorophyll content of the leaves. Mostly the chlorophyll content will be dominant in the palisade layers. The NIR reflectance can be based on the mesophyll cell structure and the cavities present in them (Campbell, 1976). There is a strong relationship between the leaf N concentration and chlorophyll content of the leaves. Higher biomass levels and leaf area can bring about higher NDVI values. This is due to high reflectance levels. The variables selected will be based on the N content. In maize, the leaf reflectance was studied at 550nm (Alchanatis et al., 2005). Some studies reported that the data collected from the remotely sensed crops can give us accurate N availability.

The growth and productivity of the crops can be determined by many variables like soil management and water/weather. The yield of the crops can be defined by several factors like biotic stresses, cultivation practices, weather, soil and crop cultivars. We can use a crop simulation approach or empirical approach or both. There are many sensors that can be classified as multispectral in nature. We can use both the visible and near IR spectrum. This has become mandatory in the modeling of yields. The remote sensing data can be synoptic, accurate and timely in nature. We can use the remote sensing data in estimating both crop growth conditions and yield rates. We can even get the yield forecasts based on the spatial scales. Leaf area index or LAI can be the area of a one sided leaf for a unit area on the ground. This can be considered as an important factor in studying the structure of the canopy.

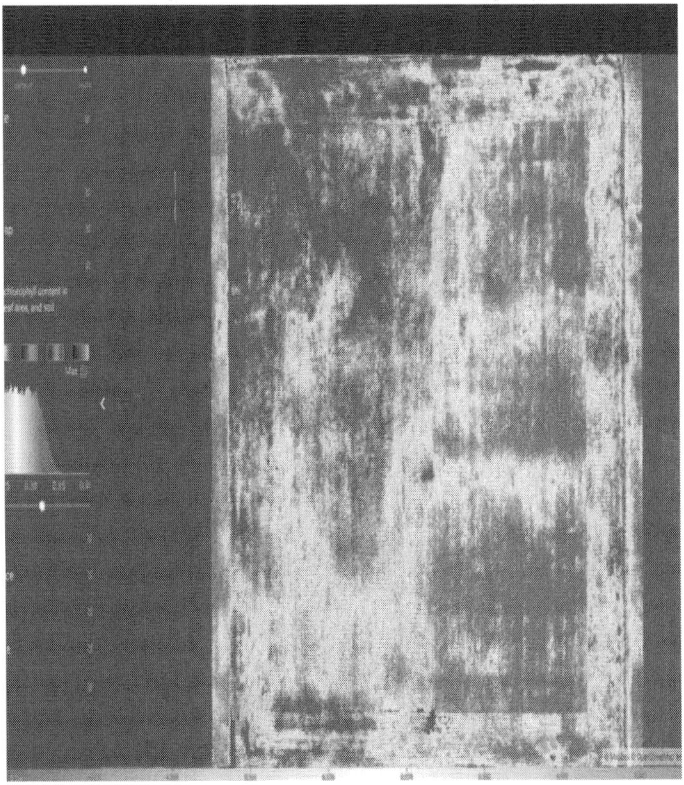

Figure 9.4: Multi-spectral data.

Source: https://www.skytechaerial.co.uk/multispectral-data-for-agriculture-2018/

The multi-spectral data (Figure 9.4) can be obtained from sensors onboard satellites like Landsat, IRS and SPOT. This data can be used in the yield modeling. The remote sensing data can be used according to our needs. We can integrate GIS data, RS data and crop simulation models at diverse spatial scales. We can use the leaf area index or LAI to measure or quantify the canopy structure. This can be dependent on the litterfall of leaves, respiration, photosynthesis, light interception and evapotranspiration. All these processes can be considered as canopy processes. Remote sensing data based on LAI can help us in many ways. We can use this data as input for several models. The inputs can be loaded into models pertaining to models like evapotranspiration, yield simulation models, crop growth models and photosynthesis. We can also use this data as input in the vegetation functioning model, biosphere functioning model, net primary productivity model and other models that can help us in measuring the crop canopy processes. These models can be applicable to larger areas. There are many models that may need data from the space-borne remote sensing devices. The data can be tested and developed according to the user's needs. We can use regression models in order to know the characteristics of the crops.

We can use statistical models that can help us in linking band radiance and LAI (Badhwar et al., 1986). We can generate the LAI-vegetation index. There are many biophysical models that can help us in this process. We can use a LUT based model or numerical model that can be used in the inversion of reflectance. We can use NOAA-AVHRR data for this purpose. In order to estimate the LAI on a global scale we need NDVI values after atmospheric correction. This approach was developed to know the accurate estimation of the global leaf area index (Gregory et al. 2003). The radiative transfer models can be used in estimating land cover. Simple NIR-Red ratio or SR can be used in determining NOAA-AVHRR (Figure 9.5). NOAA-AVHRR data can be used in sensing coniferous forests, deciduous and mixed forests and other crops. We can use MODIS data also for this purpose and TERRA platform derived data can be useful for this purpose. We can use an eight-day composite for leaf area index. The spatial resolution obtained can be 1km. This data can be incorporated in the model and we can have a look-up table. This table can be called as LUT and this can be based on LAI retrieval algorithms. The associated platform for this can be called as MODLAND. It is essential to validate these products. The validation can be essential prior to its injection in the modeling process. We can use these products in the development of LAI maps. In this process, we can use IRS-LISS III data. These maps can be produced after performing field sampling and other regression modeling approaches. Then we can get MODIS LAI product. We can measure the aerosols in the upper atmosphere using optical thickness. This can quantify water vapor. The water vapor content can be estimated along with the LAI values. The LAI values generated can be used in the conversion of the image's values into digital values. In this case, the images must be geo-referenced and the fields should be subjected to georeferencing with a high rate of accuracy.

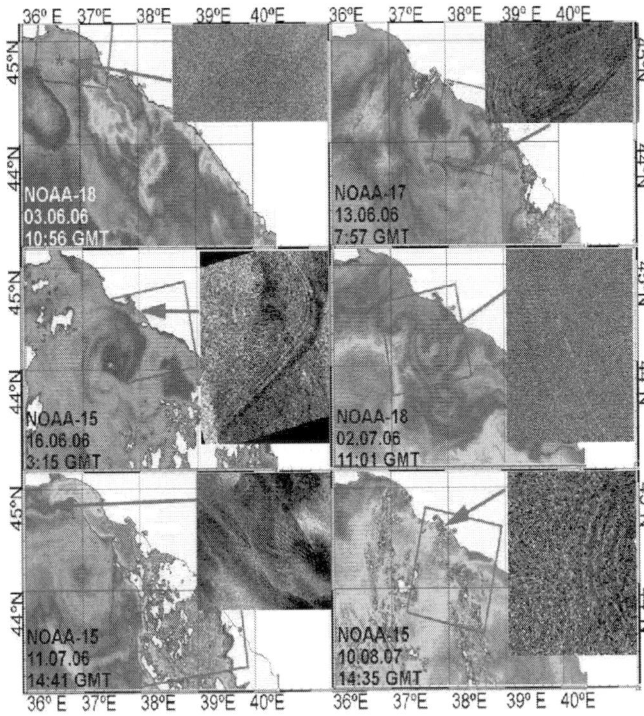

Figure 9.5: NOAA-AVHRR.

Source: https://www.researchgate.net/publication/233156095_Multi-sensor_survey_of_seasonal_variability_in_coastal_eddy_and_internal_wave_signatures_in_the_north-eastern_Black_Sea/figures?lo=1&utm_source=google&utm_medium=organic

The LAI measurement extracted from the fields should be matched with the image properties. With the help of the NDVI data obtained, we can generate empirical models. We can relate NDVI values with LAI values in order to get accurate results. This can be used in the production of LAI maps. These LAI images can be aggregated to a spatial resolution that can be equal to 1Km. We can compare this product with MODIS LAI products. This will give us a correlation between MODIS data and LISS II data for LAI. There might be some positive bias. Crop simulation models are dynamic in nature. They can be classified into state variables, rate variables and state variables. The state variables can be represented by the nitrogen content present in the soil and plant, biomass and other components at specific time scales. The driving variables can be classified as forcing functions. They can determine the effects that environmental conditions can endure on the system boundaries. These values must be monitored uninterruptedly. Some of the meteorological variables can be classified as driving variables. The state variables can be linked with the rate variables. The rate variables can be linked with the change that might occur at a certain time. This can be based on certain processes. The rate variables can include material flow and biomass flow. These processes or flows can be between the state variables. The values obtained can be based on the driving variables and state variables. It is essential to have complete knowledge of biological, chemical and physical processes that might occur whilst different crop growth stages.

Geographic information systems can be classified as an integrated system that comprises a set of tools needed for collecting/extracting, storing, retrieval, transformation/manipulation and display/presentation of spatial data of real-world for good and services determined by the user. In the GIS process, we can use computer hardware, computer software and digital data that can be shared among the end-users and research communities. The information that can be stored can be classified into two groups. The information that is spatially referenced can include points, line and polygons. These can be referenced with the GCS or geographic coordinate system. The data obtained can be stored and saved in raster and vector formats. The raster format can be called a grid-cell format. The vector format can be called an arc-node format. The second type of data that can be used is the attribute data that can be generated by GIS. This data can describe specific characteristics of any feature or spatial feature obtained from the remote sensing platforms.

The utilization of remote sensing and GIS data in the monitoring of crops can be essential. We can use this data in the many phases of the modeling process. We can use it in the preparatory phase, analysis and output phases. GIS can be used in the zonation and stratification with the aid of input layers. Some of the input layers are physiography, soil and climate. We can prepare input data that are associated with soil and water. This data can be available in many formats that can be accessible by many. Most of the GIS operations are based on raster layers. The NDVI values can be extracted from the raster layers. These values can be extracted from the administrative boundaries. In the output phase, we can expect the aggregation of GIS in order to generate meaningful data of the administered regions. We can produce maps and even we can think about the overlays. NOAA-AVHRR-NDVI data is being used by NASS or the National Agricultural Statistics Service. This data can be used in the assessment and monitoring of damage caused by drought and floods. This data can provide immense resources for analysis by cartographers and researchers. We can combine satellite data with GIS. The quality of data can be enhanced by country boundaries and overlaying of images. Raster based or grid-cell data can be used by Arc/Infor or GRID. We can produce a differential image that can help us understand the change detection over several years. We can use the overlaying option to know the effects of the crops. The precipitation data can be subjected to TIN function whilst understanding the NDVI differential image. The crop simulation models can execute input data that can be generated from the fields. The observations can be made in the form of point outputs. The simulation models can be extended to several inputs that are arranged in wider spatial domains. The basic purpose of integrating GIS and interfacing models is to perform both spatial and temporal analysis. We can even understand the behavior of the crops at regional scales. These simulation models and spatial dimension can generate temporal outputs. The GIS platforms can visualize the products generated as a result of spatial analysis. The modeling tools and GIS are in the market for years. The conceptual network has become popular these days. GIS can be linked with the display of spatially referenced objects and there will be the generation of outputs. Interpolation can be considered as a simple approach in the production of model outputs. We can use advanced strategies like overlaying and interpolation so that we can produce a database that may contain inputs for a specific model. These model outputs can be extracted to another database as decided by the user. The communication between the model and GIS can be based on identifiers. The common identifiers are polygon and grid cells. These identifiers can be used whilst the transfer of files between various models and GIS platforms. This approach may not be used in all cases. This approach cannot be useful to its full potential. There are many dependencies with the GIS formats and model. The incompatibility across the operating systems can be an important problem. There will be no complete utility of GIS capabilities. We can combine the data with the model results. Most of the models can be configured according to GIS platforms. In the GIS package, there will be libraries, interface programs and macro languages. There is a need for complex programming in this case. We can link data management with GIS data. We can integrate one system to another via GIS platform bases also. There is a need for experts and a lot of effort in this process. In the linking process, there will be a file exchange between GIS and the model. Both GIS and model have a user interface. The user interface will be utilized by the end-users so as to achieve the desired results. In the combining process, there will be a file exchange between GIS and model. In between the GIS and model, we will have a user interface. This user interface will be used by the user communities. In the integrating step, we can use a common software system. GIS will be used in the pre-processing stage. After this stage, the data will be sent to the model. There will be post-processing stage. After this stage, we can inject the data into GIS after quality control. Then we can create a platform for the user interface. This interface can be shared among the user communities. This organizational structure was proposed by Hartkamp et al., 1999. This GIS interface can combine most of the crop management practices according to space and time and then we can evaluate crop production. (Figure 9.6)

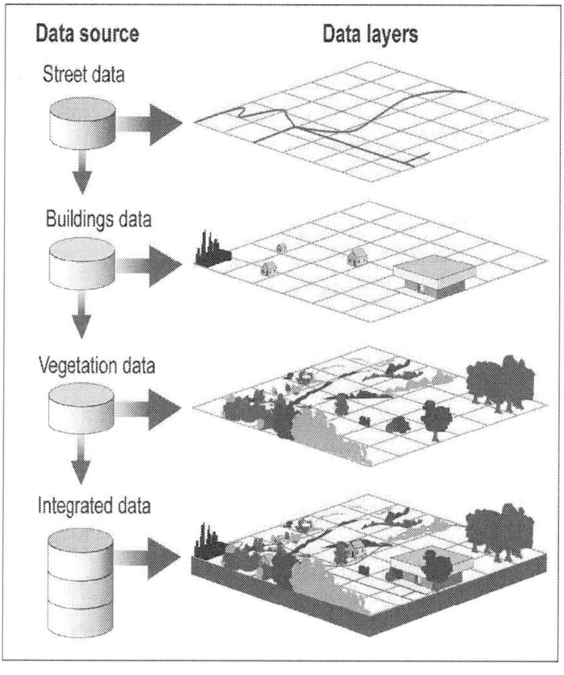

Figure 9.6: GIS.

Source: https://www.nationalgeographic.org/encyclopedia/geographic-information-system-gis/

The agricultural decision support system was called as Agricultural and Environmental Geographic information System or AEGIS. AEGIS system uses DSSAT capabilities. These capabilities are present in ARC/INFO GIS. This can be used in production analysis and regional planning. This system allows us to combine all crop management practices. AEGIS was upgraded into AEGIS for windows and it was called as AEGIS/WIN. This system was written in Avenue and it was a type of macro scripting language. This program can link Arc View-2 which is a geographical mapping tool with DSSAT-3. Spatial analysis software was developed by Thornton et al., 1997. This software can link GIS software and crop models. The spatial analysis software can be classified into 2 modules. The geostatistical module can be used in the interpolation process. This module can be to interpolate maps. We can produce surface maps that represent the network of the points that represent data. The utility module can be used to measure the statistics of economic outputs and agronomic outputs. This can be achieved by model simulations. There will be the production of polygons from the maps. SPATIAL-EPIC was developed and that was linked with Arc/Info. Many reviews were made with GIS and modeling. Arc/Info and WOFOST models were used in the yield forecasting in the European countries.

There are many databases pertaining to yield statistics, crop, weather and soil properties. The analysis part comprises three modules. The three modules are agrometeorological modules, statistical modules and crop growth modules. In the case of the meteorological module, we can process meteorological data on daily basis. The data can be obtained on a real-time basis. We can use a regular grid in this case and it will cover 50 x 50 km. This can be used in the assessment of alarming conditions and crop growth models. In the crop growth module, we can use a dynamic simulation model like WOFOST. In this case, the crop growth patterns can be calculated. We can use crop indicators. We will have two production levels namely water-limited level and potential level. WOFOST can be executed on daily basis and so we call it a simulation unit. This can be an association of crop, soil and weather unit.

In the case of the statistical module, the crop indicators can be calculated by WOFOST. The crop indicators are linked with dry weight storage organs and ground dry weight. In this analysis, we can use yield statistics obtained from historical data. This can be done with regression analysis. The regression analysis can be associated with the time-trend. We need to use 15 years of historical data and then simulate it. The regression equations that were generated for every crop for every region can give us an accurate forecast. The CGMS can produce three kinds of output. The maps can be produced based on the weather variables on daily basis. The grid size will be 50 x 50 km. This can give us an idea about abnormalities like frost and drought. The maps can be generated with the agricultural quality indicators and this can also be based on the comparison studies with the crop indicators and this time it is long-term. Here, we can get both tables and maps that represent yield forecasts. There are five important methods with which the remote sensing data can be integrated with the models.

They are 1. Using a driving variable that can be generated from remote sensing data in the model. 2. We need to update the state variable that is a part of a model. In this case, we can consider LAI that can be derived from remote sensing data. This can be called a forcing strategy. 3. We need to re-initialize the models. We can adjust the initial conditions so that the simulation can get suited to our argument. All these inputs can be derived from the remote sensing data. 4. We need to use a corrective method. This can develop the relationship between the error generated in the final yield statistics and error produced from the intermediate variables that might be derived from the remote sensing instruments. We can use this relationship in cases where the final yield statistics are unknown.

The simulation models can have driving variables. The inputs in this case can be weather data that is recorded on daily basis. Some of the minimal subsets that can be used as inputs are wind speed, relative humidity, solar radiation, minimum and maximum temperatures. This approach comes with some drawbacks. In the case of remote sensing derived data, the unavailability of data, cloud cover and other factors can be considered as drawbacks in this approach. Among all the variables available, the absorbed PAR, solar radiation and rainfall can be important.

In the case of forcing strategy, we need to update state variables that are to be used in the modeling process of remote sensing data. In this case, LAI can be considered as a state variable that is to be updated. The forcing process can be performed on the same day of data acquisition. We can produce LAI profiles with parametric models (Wang et al., 2016). In the case of the re-initialization method, we can rely on the model performance and it is based on the state variables at the initial stage. We can manipulate the initial conditions wherein the state variables are generated. This will lower the deviations between the radiometric signal and derived state variables and eventually simulation. In the case of a re-calibration strategy, we assume that the model in question is inadequate and mostly it may need re-calibration. This can be done by limiting error propagation. The re-parameterization and re-initialization sectors in crop models may need radiometric information. We may not need derived canopy parameters in this case.

In the crop growth monitoring system or CGMS, there will be linkages or associations between inputs, GIS and databases. The inputs that can be used are grid size, area bounds, administrative boundary map, classified remote sensing data, weather data on daily basis and basically points, weather constants, soil maps, soil organic content and crop management practices like fertilizer input, irrigation, sowing date and cultivars. In the GIS portion, we can use the model grid, boundary vectors of the study area, crop mask, weather surfaces created by interpolation, Thiessen polygons, soil texture classes, soil depth and soil organic content surface maps with interpolation and crop yield map. In the relational database sector, we can use grid number (GN), geographic area comprising of latitude and longitude values, district code, crop fraction area, weather files and constants, soil depth with GN, soil organic content with GN, cultivar and GN, sowing date, crop yield with GN. With this data, we can get aggregated crop yields. In the simulation model sector, we can use model outputs and others. There are models that can give us yield predictions with the aid of remote sensing data. They can be classified into spectral models, albedo models and thermal models (Tucker et al., 1997). Three models were framed to forecast the yield and growth of the crops. The models can be classified into empirical regression models, biomass production models and stress-degree day models. In the case of meteorological models can be used to predict the yields and it is dependent on precipitation and temperature variables.

These variables can be based on crop yields. The data can be extracted from the meteorological stations. We can use satellite-based measurements to get the meteorological data. We can use both precipitation and temperature data separately or in conjunction. In this case, we can use both monthly and daily variables. Most of the meteorological variables can be based on simple regression algorithms. These regression models can be classified into three primary methods. They are the time series method, variation in space and time method and space variations method (Lobell et al., 2009). We can use a statistical approach to predict crops and we can study it as a function of rainfall variable. If the crop to be studied is rain-fed, we can consider rainfall as an essential factor. In this case, we can use rainfall as an important variable in the prediction of crop yields and growth. We can use rainfall-dependent models to measure potential yields. In some cases, these models might not be useful as they give us overestimates of the yields. In these models, there is no specific consideration of soil water storage, drainage, runoff and distribution of rainfall. We can use the ratio of potential ET and actual ET to forecast the crop yields (Fitzpatrick et al., 1989).

Whilst using agrometeorological models, we need to consider yield components, energy balance, soil moisture and temperature. The agrometeorological models can be linked with the remote sensing models and methods. It is essential to predict the crop yields in limiting growing conditions. This is often hard to obtain. We can use physiological variables and remote sensing data to get real-time information about the growth conditions of the crops. PAR can be classified as an essential physiological variable. We can link the leaf area index or LAI with the data derived from the remote sensing devices. The greenness and Vegetation indices can be linearly linked with the PAR absorption rates. We can link the canopy temperature with the yield predictions. It is essential to evaluate the driving variables. The evaluation technique can be simple whilst the usage of remote sensing information. The remote observations may be associated with the growing seasons. We can acquire data according to the feasibility of the location and measuring capabilities. If the data acquired is less frequent, then we can use interpolation techniques. With the advanced interpolation techniques, we can get the daily values of the variables available. It is essential that the remotely sensed data must be used in every stage of this process and that is not possible.

It is important to update the state variables present in the model. The state variables can be simulated on daily basis. If there is available remote sensing data, we can use this as basic data and can do the analysis needed as per the requirements. There are many crop models that can use LAI values. These LAI values can be used to predict the accumulation of biomass in the fields. The values pertaining to biomass ad LAI will be simulated via model. According to the remote sensing data availability, these variables will be updated. Remote sensing data may have errors to some extent. It is important that the SVs should be updated regularly according to remote sensing data acquisition (Cammarano, 2014). We can reinitialize the SVs. In this technique, we can modify the initial conditions in the model and it will be according to the remote sensing data. We can do re-parameterize the models. The basic need for the integration of the CSMs and RS for crop management between-season and in-season can be based on remote sensing observations, soil-crop weather situation and crop simulation.

The drought monitoring situation can be forecasted with the aid of seasonal rainfall data. Food and Agriculture Organization or FAO has developed an index called as WRSI or Water requirement satisfaction index (Frere et al., 1986). This model can be used to forecast the effects of precipitation in crop production, especially in the rain-fed areas. This tool can use decadal information pertaining to precipitation. It also uses water holding capacity values of the soils. We can calculate the demand for water with the help of potential ET values. The supply can be estimated from soil water and rainfall events. For African countries, the WRSI tool was developed. We can use point WRSI and RS NDVI to obtain the spatial maps. There are several predictive approaches that can integrate both climate data and remote sensing data. The approaches can be on short, medium and long-term basis. These approaches can be used in simulating agricultural productivity. The magnitude of reflectance obtained from vegetation can be higher than the inorganic materials. In this case, vegetation can be brighter when we see it under the NIR spectrum. The classification of crops can be a laborious process and it is always difficult to demarcate the crop types. The classification system or process can be based on backscatter and reflectance levels. The primary problem in this case is that the reflectance values and backscattering values can give us less discrete data with which we may not differentiate crop types properly and accurately. In the canopy, there will be multiple scattering. This property can be used in the demarcation of crops. This demarcation can be based on remote sensing with radar. Among the radar remote sensing devices, SAR or synthetic aperture radar can be an important and reliable option in monitoring crops. It is essential to use appropriate backscatter signatures and reflectance values and they should be well defined so that demarcation will be perfect. We can integrate radar imagery with optical sensing techniques to get accurate crop demarcation. The SAR sensors can be useful in closing the gap that might have occurred in the optical sensing even in the presence of cloud cover. Differences in crop cultivars, variable sowing seasons, shapes and agricultural plots can be considered as important factors while identifying the errors.

Chapter 10

Unmanned Aerial Vehicles

CONTENTS

10.1 INTRODUCTION

One of the novel developments in the agriculture sector is the production of unmanned aerial vehicles (Figure 10.1). They are called drones and they are small. These drones can be controlled remotely and there is no need for a human pilot on-board.

Figure 10.1: Unmanned Aerial Vehicles.

Source: https://en.wikipedia.org/wiki/Regulation_of_unmanned_aerial_vehicles

They can be used in agriculture to support planning and in the collection of spatial data. Drones can provide us valuable data that can help in making policies and decisions. They can be used in other fields like the military, disaster management, etc. The utility of drones in agriculture can be used in the production of crops, wildlife conservation, fisheries, forestry, risk reduction, early warning systems, etc. Precision farming can integrate imaging technology with the sensor data on a real-time basis and this will improve mapping in the field. The data collected from the drones can be raw data and this should be processed so that it can be used in analytical agricultural models.

Figure 10.2: Drones.

Source: https://www.popularmechanics.com/technology/gadgets/g2983/the-best-drones-review/

Drones (Figure 10.2) can be used in determining the soil health condition, weather analysis, yield prediction, fertilizer application and irrigation schedules. The data obtained from drones can be linked with data sources and analytical techniques so that we can get reliable information that can be useful. Drones can be used to know the variable rates at which the herbicides, fertilizers and pesticides be used on the crop plants. We can use hyper-spectral and multi-spectral satellite imagery to derive NDVI values followed by maps. These values can help us demarcate crop stages, stresses, forest, grass and soil. Many researchers have observed a strong connection between NDVI data and crop yields (Huang et al., 2012). It is essential to track crop growth stages and this will help us to estimate the crop yields accurately. With these estimations, we can address the issue soon before it is too late for an action to be taken. The drones can be equipped with hyperspectral, multispectral and infrared sensors. These drones can be used in the analysis of soil conditions and crop health precisely. NDVI data can be integrated with the canopy-chlorophyll content index, crop-water stress index to get agricultural mapping tools so as to get crop health status. Drones can be used in estimating agriculture insurance and legal metrology. The drone imagery can be utilized in the estimation of losses accurately.

Rice crops can be considered as an important crop with which we can use integrated remote sensing techniques to get the yield statistics. The satellite-based rice monitoring system can be called an SRM system. This system integrates ICT tools, crop models and remote sensing tools. This system can produce accurate information pertaining to the growth stages of rice crops. It also gives us statistics about biotic and abiotic stresses. These integrated systems like RIICE can use Amazon web services or AWS, unmanned aerial vehicles or UAV, smartphones, web geographic information system (GIS), crop modeling and remote sensing. SRM system assisted in the production of sustainable technologies. The SRM system can be based on three software modules. They are ORYZA crop growth model, Rice yield estimation system or Rice-YES and MAPscape-Rice.

The Earth observation data will be loaded into MAPscape-RICE in order to process SAR data. We can generate data from administrative units, crop practices and crop calendars. We can use in situ point data of leaf area index as input. After processing SAR data, we can validate drought damage, flood damage, seasonal dynamics, leaf area index, pre-season and seasonal area statistics. Then the processed data will be subjected to a rice yield processing module via Rice-YES and ORYZA. In this module, we can use management data, phenological data, soil and meteorological data. Post-processing at this module, we can validate the yield estimation and yield forecast. This will help us estimate both production and production loss.

Drones can be used in field analysis and soil analysis. It is essential to produce three dimensional soil maps. We can plan sowing times and nutrient status of the soils using these maps. This will drastically lower the costs incurred whilst planting. The drones can scan all the field area and can help us spray the liquids in exact amounts. We can modulate distance and can manipulate the spraying times along with the coverage. We can do crop monitoring with the help of time-series animations. These animations can help us know the accurate developmental stages. This will also help us to know the plant health and estimate production problems. It can be used in crop management strategies. If drones are equipped with NIR and visible sensors, we can even identify the portions of the fields that need improvement according to irrigation resources. We can use NIR and visible sensors in scanning all the properties of the field crops and so we can assess crop health.

There are many problems with UAV implementation. It is essential that we need quality software, data, acceptability, flight range and time, etc. UAV implementation demands quality software so that the final image can be processed after each sortie. Some nations may not allow drones easily due to legal complications. The farmers must be in pace with technological awareness so as to accept drones. This can be stated as one of the hard hurdles. Drones come with flight ranges that are too short and so the acreage to be covered will be limited. It is to mention that the drones with longer flight range may be expensive and farmers may not afford them easily. The initial costs of these drones will be high and this may not be feasible for small and marginal farmers unless subsidies are provided by the governments. Drones flight at low levels, but they may share airspace along with other aerial vehicles and this needs more approvals. The interference must be avoided at all costs and so the drone's mobility is limited to certain places. Connectivity issues can be the main problem. The farmlands may not be well connected.

This technology can offer many advantages across the crop value chains. These chains will extend from farmers to the consumers. The farmers can be considered as direct beneficiaries and they are classified as end-users of technology. They can get real-time information about the crops. With the data obtained they can make some informed decisions. With this data in hand, farmers can plan an entire crop cycle and manage their operations with limited damage. In this case, the initial installation costs can be high. The damages that can be incurred in absence of drones might outplay their presence with luxuriant crop growth and economics. This can be classified as a feasible technology that can be adopted in the near future. Damage to the crop plants can be reduced with the use of data and this will help us define the health indices of the crop plants. This can considerably enhance the income of the farmers. We need to optimize the farm operations and this can help us in enhancing the stocks. The mechanized operations can be optimized using the drone technology. Monitoring of irrigation practices, soil mapping and crop health can be easier with drones. Drone technologies can be associated with cloud-based data analytics. This can be used in the detection of pest infestation, nutrient stress and water resources. This will help the farmers to get data from handheld devices. Advanced algorithms can be used to estimate population, species identification and localization.

Drone technology can be used to estimate the growth stages of a plant. There are several deep learning algorithms that can be employed in the identification of species. We can even understand the infrastructure that is common for two features or shared by two features. We can evaluate the proximity levels.

10.2 EMERGING TECHNOLOGIES

There are several emerging technologies that can help us in advanced applications like precision agriculture and smart farming. Internet of Things or IoT can be classified as advanced technology. These technologies can help us in acquiring environmental data on a real-time basis. The devices equipped with IoT can be used in crop management applications. This can be performed with the extraction of both spatial and temporal data or images. IoT technologies can improve the production rates drastically. Remote sensing can be classified as an essential and popular technology for smart farming and precision agriculture. This tool can be used in monitoring the cultivated fields. This technique can monitor crops and vegetation parameters with the aid of wavelengths. UAV-based remote sensing devices are more advanced and improved precision agriculture. The UAVs have extended possibilities in the process of acquiring field data. This process has become faster and comfortable. The costs incurred in the extraction of the data has reduced drastically. IoT with UAV can be classified as the next stage of development in precision agriculture. These devices can fly easily at low altitudes and can give us ultra-high resolution images.

The resolution of these images can be in centimeters. This will enhance the efficiency of the monitoring systems. The UAVs can have a high temporal resolution. These devices are simpler and can be cheaper if subsidized by governments. They can cover larger areas in less time and we can classify these devices as non-destructive. In the initial days, the drones were used for military purposes and then the application of drones was extended to agriculture and other scientific needs. The primary goal of these UAVs was to derive the data from image-based products. Most of the UAVs that are being used in agriculture can be intelligence dependent products. These products use algorithms that can process images systematically and easily. The outputs generated from these intelligent processing algorithms can help the farmers in the decision making process. The popular application of UAVs is crop spraying, irrigation management, disease detection and vegetation health monitoring, yield estimation and vegetation growth monitoring, management and mapping of weeds.

The weeds in the croplands can be considered as undesirable as they generate competition whilst nutrient uptake and recycling. This can cause many problems in the crops. More competition will get generated for both water and space. This will reduce crop yields. It is essential to use herbicides to control these weeds. This will produce herbicide-resistant weeds and this can be considered an important problem in the crops. Over usage of the herbicides can enhance the pollution levels in the environment and can result in eutrophication. Site-specific weed management or SSWM was modulated to address the weed-related problems. In this system, weeds can be classified according to the zones that can be manageable. It is to mention that some of these management zones will have a weed population. That means few spots of the cropland comprise of weeds that can be detrimental to the crops. In order to control these weeds, we need an accurate weed cover map. This will help farmers to determine the precise spraying limit of the herbicides. The UAVs can capture images and from them, we can derive valuable data of the entire field. The weed cover map generated can exhibit spots where we need to spray more chemicals. We can grade the number of herbicides to be sprayed. We can use UAVs frequently to monitor yield estimates and vegetation growth. These drones can be used in observing crop health if weather conditions are calm and conducive for the sorties. We can easily detect the diseases and the changes that might have appeared on the plants in the form of symptoms. We can observe the variations induced by biochemical and biophysical characteristics in presence of a disease. The data processing in UAV can be mostly image processing and the outputs can be useful in measuring the biomass of the crops. We can easily detect the symptoms in the early stages of the disease. We can use UAVs in two phases. We can detect the initial infection stage and this can be achieved by grasping enough plant health information.

We can map the infection size. We can employ a targeted spraying technique with UAVs and this will reduce the costs. The irrigation management of crops can be considered as an important application in UAV technologies. This technology can be used in precision agriculture. Most of the crops cultivated all over the world are based on irrigation systems. It is obvious that precision agriculture must be improved in order to get better results. We can use precision agriculture to enhance water use efficiency in the croplands. The UAVs that are being used in precision agriculture can comprise some of the essential elements. We need UAVs and they need to be operated remotely or in some cases, we can prefer autonomously operated UAVs. We need a ground control station or GCS. This can be classified as a computer that can effectively communicate with the UAV directly and uninterruptedly. We can easily regulate the movements of UAVs. The users can obtain data related to the sorties made and the sensor information that is the wavelengths being used. At the ground control station, we will be having software to process data by the operator. This processed data can be used in crop monitoring. It is important to have a separate UAV control system or in short, we call it UAV cs. This can be used in controlling the drones as and when required. This can be arranged as an in-built system or in the form of remote control. We can keep GPS as an integral entity in the drones. In this system, we can have a flight control system and autopilot system. These systems can control all the movements of the drones. The sensors onboard an UAV can be important. In the case of fixed wing drones, there will be a runway for a specific distance or even we can use a catapult. This type of UAVs can provide us a high level of endurance and we can aim for more heights as and when needed. These UAVs can be used to cover most of the areas. This type of UAVs can be expensive and so we need to make sure that the expenditures are justified. We can use solar-powered UAVs as they can offer us extended flight times and this is due to the fact that they use natural sunlight. In the case of extended flight operations, we need to use this type of UAVs. We can call extended flight operations as long-endurance operations. There is another type of UAVs called as rotary-wing UAVs. They can be called rotorcrafts. They can be used in vertical take-off and landing capabilities or VTOL. This helps us in flying the drone steadily and with excellent maneuvering abilities. We can utilize these features in a variety of missions. The main problem with these drones is that they cannot stay on air afloat for long periods and are not capable of very high-speed operations. In precision agriculture, we can use this type of drone. Blimps are another type of UAV that can be lighter than air with a high rate of endurance. This type of UAVs can be operated at low speeds. This type of UAVs can be larger than other UAVs. They can be afloat even when power is less or unavailable. This type of UAVs can be classified as safe.

The sensing systems in UAVs can be powerful and they can be linked with IoT based technologies. We can use the sensors to capture images. These images can have high spatial and temporal resolutions. We can easily detect various characteristics of crop types. The sensors should be lightweight and they need to consume less power. The sensor capabilities should be upgraded according to the needs. The sensors that can be used in UAVs can be classified as thermal sensors, hyperspectral sensors, multispectral sensors and visible light sensors or RGB sensors. Every sensor onboard the UAV can capture a specific type of image that can help us in determining the texture of vegetation and color. We can also define crop outlines. We can process the data obtained from these sensors to evaluate crop characteristics, soil moisture, vegetation health and plant biomass. The RGB sensors can be used in the UAV systems and they are very useful in precision agriculture. The costs incurred whilst operating these UAVs can be less. The information captured by these sensors can be subjected to simple processing so that we can easily extract the attributes. We can get images on cloudy and sunny days at certain time frames. Some of the vegetation parameters may not be studied with RGB sensors. This is due to the fact that extensive analysis of vegetation characteristics needs spectral information beyond the visible spectrum. We can use RGB sensors along with other sensors.

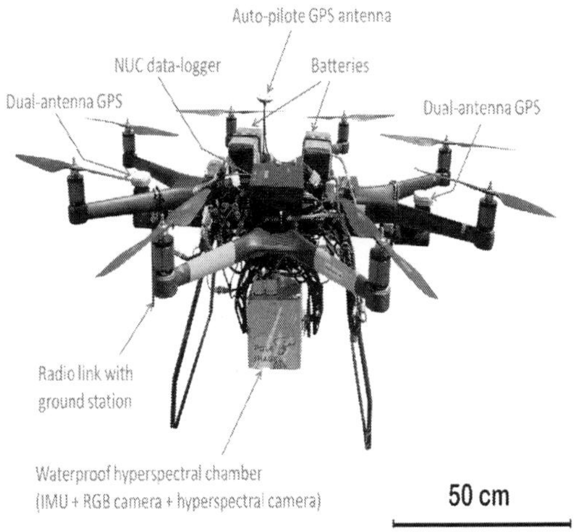

Figure 10.3: Hyperspectral imaging sensors.

Source: https://www.popularmechanics.com/technology/gadgets/g2983/the-best-drones-review/

We can use multispectral imaging sensors and hyperspectral imaging sensors (Figure 10.3). The UAVs can be useful in obtaining information about vegetation. These sensors can provide us spectral absorption and reflectance values with high accuracy. This can be obtained in several bands. The derived spectral data can be used in evaluating both physical and biological characteristics. We can easily demarcate the unhealthy portion of crops using the images generated from these drones or UAVs. We can use the red channel as this wavelength can be absorbed by chlorophyll molecules. In this case, the NIR radiation will be reflected. The NIR and red wavelengths can be used in producing vegetation indices. The multispectral and hyperspectral images may need complex processing methods. The extraction of information from these images can be expensive. We need to use techniques like image enhancement, image fusion, geometric correction and radiometric correction. The number of bands can differentiate the hyperspectral and multispectral images. In the case of multispectral sensors, we can get data from 5 to 12 channels. In the case of hyperspectral images, we can get data through a thousand bands or even hundreds. The bandwidth will be narrow in this case. The multispectral sensors can be used more frequently than hyperspectral images as the former are cheaper. The hyperspectral images can help us understand several crop characteristics. This will help us in phenotyping research. The thermal infrared sensors can be used to know the temperature of the objects of our interest and we can produce an image that can yield this information though there are no visible properties readily seen on the image. Thermal cameras can be used in the form of infrared sensors. We can install an optical lens to observe and record infrared energy that is reflected from the crop canopy. All the objects that are warmer than absolute zero can emit radiation. This radiation can fall in the infrared range. We can get longer wave IR and middle wave IR reflections. These wavelength ranges can be used in extracting the temperature attributes. These thermal sensors on-board can detect the radiations and give us grayscale images. These images can represent heat levels generated due to reflection from the objects per se. Some of the images generated from these sensors can be colored images. In some cases, the objects that are warm can be represented in yellow and those that are cooler can be represented in blue color. We can use these sensors in diverse applications. We can use these sensors in irrigation management practices. These sensors cannot be used in precision agriculture.

We get many overlapping mages of the crops. These sensors can capture front overlap images and side overlapping images. These images can be used in the production of both orthophotos and 3D models of the crops. The overlap rate will be based on the crop types to be sensed. The allowed side overlap can be between 40 to 95% and the front overlap can be between 60% to 95%. The altitude of these UAV sorties will be based on our requirement for spatial and temporal accuracy. In general, the spatial resolution can be about 0.5cm per pixel and this can be 10cm per pixel also in some cases. We can modify the RGB sensors. We can obtain the Red edge band or RE band and NIR band with the RGB sensors after necessary modifications. This can help us to prevent higher costs incurred in multispectral cameras. We can replace optical filters and this will help us to retrieve images in the NIR channel. This will give us a hybrid sensor called as NIR-RGB sensor. In this case, the visible channels may not be useful so that we can use other RGB sensors.

Some of the crop features that can be extracted with UAVs are vegetation color, moisture content, nitrogen status , biomass and other vegetation related features. We can get vegetation indices, size and shape of plants, spatial position, temperature and spectral behavior of the chlorophyll pigment (Varela et al., 2011). The electrical conductivity, temperature and moisture content can be observed with UAVs.

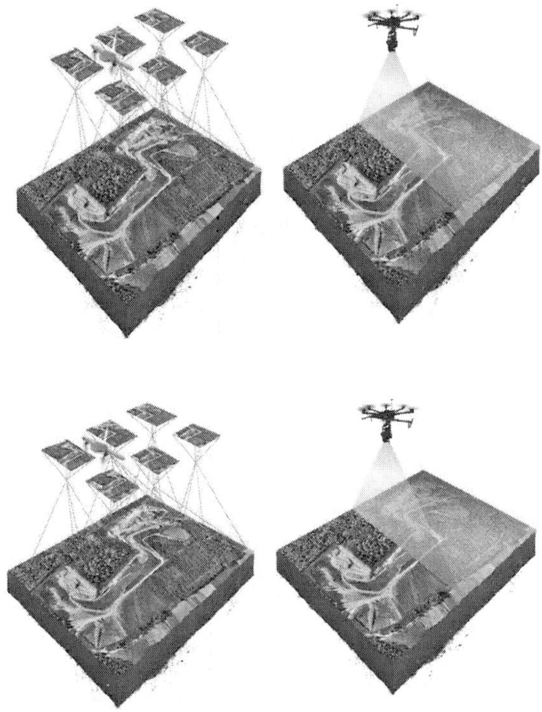

Figure 10.4: Photogrammetry.

Source: https://wingtra.com/drone-photogrammetry-vs-lidar/

We can use photogrammetry (Figure 10.4) techniques to reconstruct the scenes. This can be done by using many overlapping images. Photogrammetric techniques can be used to analyze 2D data. We can establish a geometric relationship between the objects and 3D models.

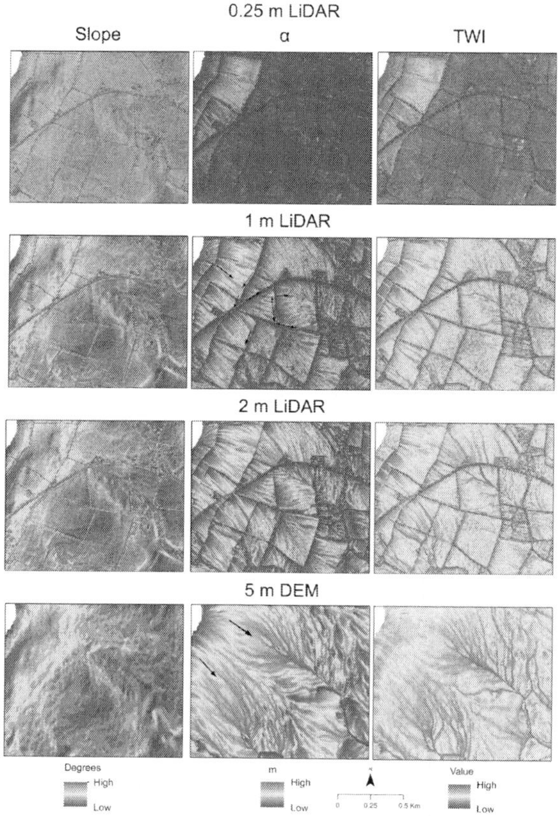

Figure 10.5: Digital elevation models.

Source: https://www.sciencedirect.com/science/article/pii/S0303243416301519

The 3D models require 2 overlapping images of any object and scene and this must be captured or acquired from at least two viewpoints. These techniques can be used in the production of terrain models and digital surface models (Sam et al., 2013). We can also produce orthophotos using these images. We can use data obtained from UAVs that are being operated at low altitudes for the production of the 3D models. This will give us a higher spatial resolution. We can construct digital elevation models (Figure 10.5) or DEMs of the crops. We can produce orthophotos in the form of orthomosaics. These two features can be used in understanding the characteristics of density, canopy, height and vegetation of the crops. These models can use RGB imagery.

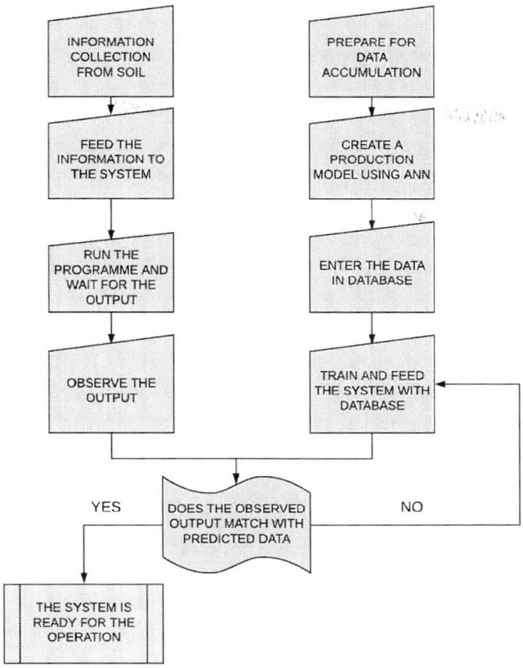

Figure 10.6: Machine learning methods.

Source: https://www.sciencedirect.com/science/article/pii/S2589721719300182

Machine learning methods (Figure 10.6) are becoming popular these days. They can be called ML methods. These methods can be used to process data. This can be used in both identification and prediction. We can use NL methods in precision agriculture. We can use these methods to estimate the growth rates, detect diseases and demarcate the objects with the aid of images generated by UAVs. We can calculate the vegetation indices or Vis. The vegetation indices can be considered as important in the remote sensing applications pertaining to precision agriculture. They may need transformations and mathematical combinations of spectral bands to get the characteristics of vegetation. This will reduce the errors generated by the external factors. Some of the software that can be used are Adobe Photoshop, Agisoft Photoscan, QGIS, MATLAB and Pix4D. We can correct distortion and can perform image processing methods with the help of Adobe Photoshop.

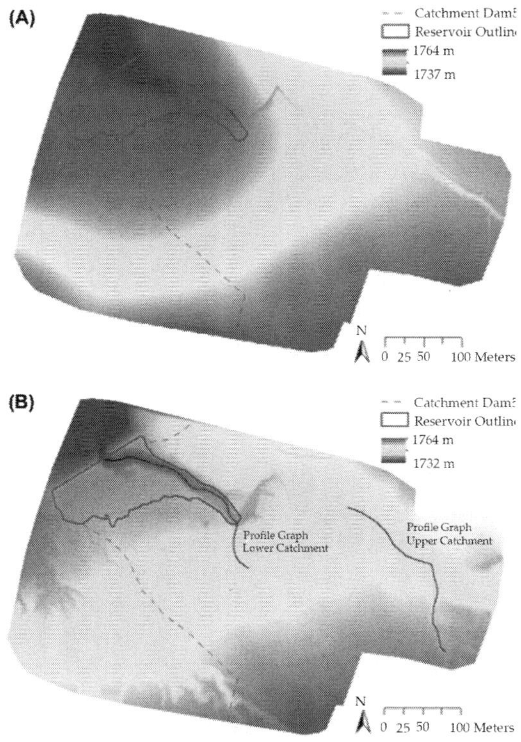

Figure 10.7: Digital terrain model.

Source: https://www.sciencedirect.com/topics/earth-and-planetary-sciences/digital-terrain-model

The digital terrain model (Figure 10.7) or DTM can be used to represent the altitude. DTMs don't consider natural and artificial objects that exist on the surface. In other words, DTM represents bare earth (Kubad et al., 2000). The digital surface model or DSM can be used to represent the altitude of the surface and this can be the top of a building or treetop, etc. The elevation model that can be generated with DSM can include elevation of bare earth along with the natural and artificial objects. The digital elevation models produced can be used in the extraction of the information from the crop orthomosaics. (Figure 10.8)

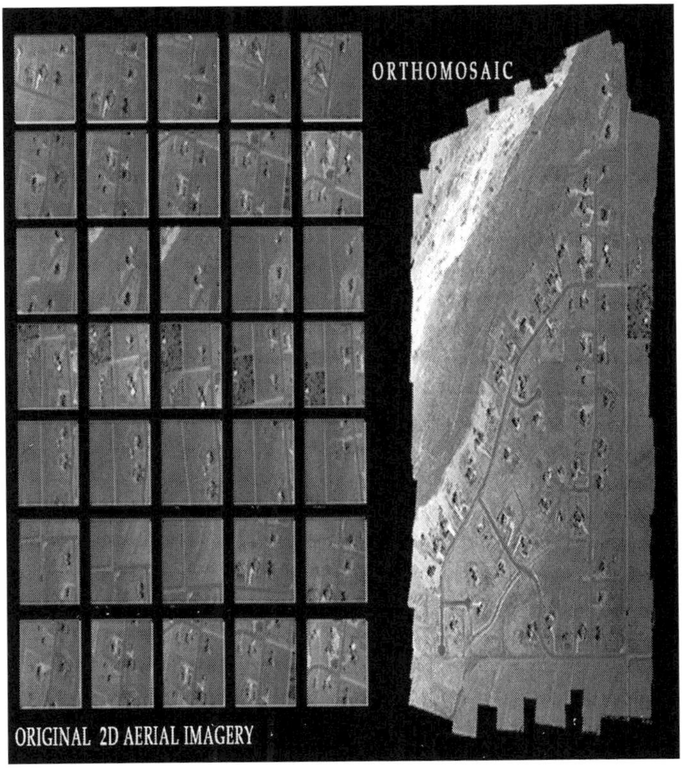

Figure 10.8: Orthomosaics.

Source: https://medium.com/new-farmer/what-is-an-orthomosaic-photo-11140c0601df

Orthophoto can be any aerial photograph that was orthorectified or geometrically rectified. In this case, the scale of the image can be uniform. In the case of crop orthophotos, we can get the exact distances and 3D characteristics. In the photogrammetry technique, we need to produce a point cloud and it is from the 3D surface. We can integrate all the objects into a DEM. With the aid of DEMs, we can produce orthophotos. The popular algorithm that can be used in this process is SfM or Structure from Motion. SfM needs no information about environmental settings nor camera parameters. These techniques can be used in many applications.

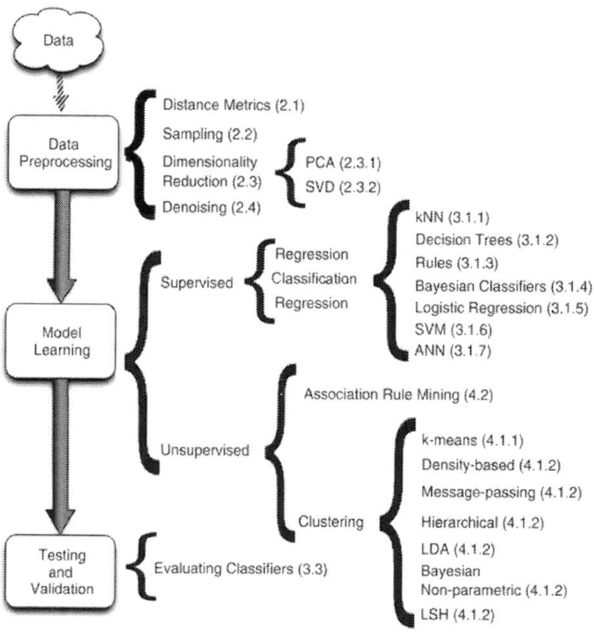

Figure 10.9: Data mining methods.

Source: https://link.springer.com/chapter/10.1007/978-1-4899-7637-6_7

In precision agriculture, we can use data mining methods (Figure 10.9) and machine learning methods to assess the information generated from the sensors onboard UAVs. The data obtained from these UAVs will be large and needs to be stored in storage devices with higher capacity. The data generated from these UAVs can be enhanced via processing algorithms. We can use both supervised and unsupervised learning techniques in regression, classification and clustering methods. Regression methods can be classified as one of the popular techniques that can be useful in all the UAV applications. This data can be used for many purposes. With regression methods, we can easily estimate the vegetation indices and mostly it will be through RGB images. With these regression results, we can get quality outputs that can offer us many benefits with the data. We can use regression to investigate the correlation of these Vis and other features like biomass, LAI and nitrogen. In this case, we can use simple regression and multiple regression along with non-linear methods for this purpose. We can use RF or Random Forests in order to get better results. In this case, we can set the coefficient of determination as less than 0.79. We can use these regression methods to predict the water status of the crops. This can be done with RGB images, thermal sensors and multispectral images. We can also use Artificial Neural Networks to yield better results even if the coefficient of determination is 0.87.

The classification methods can be best if used in weed mapping (Michaela et al., 2016). We can use these methods to detect crop diseases. We can use the Random Forest algorithm and Artificial Neural Networks or ANNs family to get better results. These algorithms can utilize RGB colors, spectral information and intensity. We can use neighboring pixels whilst using these algorithms. These classification algorithms can utilize vegetation indices to get accurate results. The ANNs can be used to get 99% accuracy whilst weed mapping. The method accuracy can be based on the crop type to be monitored. CNNs or Convolutional Neural Networks are being used by many these days for vegetation mapping. This can be classified as a deep learning algorithm. The CNNs can be used in the identification of the objects, especially with large datasets.

Deep learning (Figure 10.10) or DL methods can be used in precision agriculture and it is gaining popularity as we speak due to its accuracy and applications. The DL methods can bring more complexity whilst enhancing ML capabilities. These techniques can transform data in hierarchical methodology and there will be many layers of abstraction. The object-based image analysis or OBIA can be considered as a machine learning method that can be used in precision agriculture. This method or application can be used in demarcating objects derived from UAVs. Most of the traditional classification methods are pixel-based, in the case of OBIA, the small pixels will be grouped together in the form of vector objects. Due to high-resolution data generated from UAVs, the traditional classifications have become obsolete and less effective. This can be due to a change in the relationship between the object size and pixel size. Due to this, the classification methods have evolved into object-oriented classification methods. In these methods, we can apply segmentation methods. In this process, we can divide the pixel as homogeneous groups or segments. These objects or segments can be arranged in the form of classes. This classification will be based on the textural, spectral and geometric characteristics.

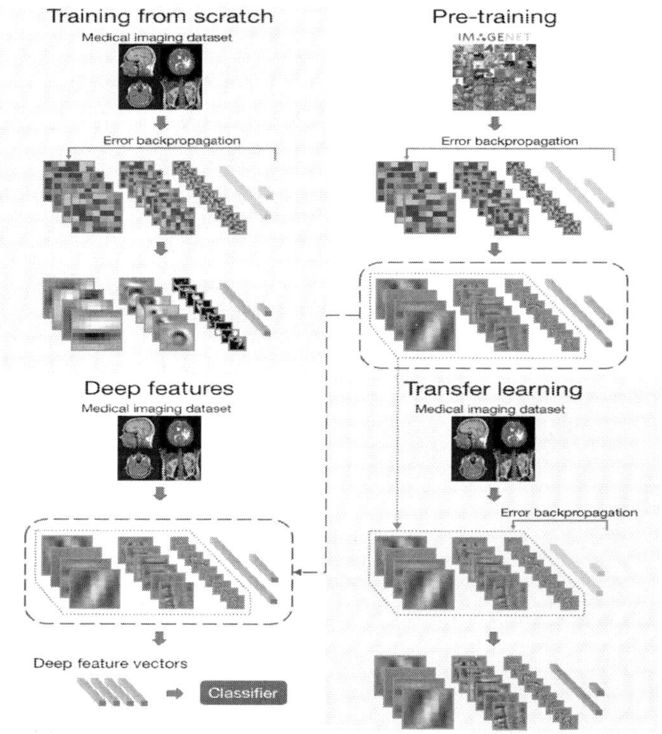

Figure 10.10: Deep learning methods.

Source: https://mateuszbuda.github.io/2018/02/17/methods.html

The OBIA can be classified into two phases. The first phase is the image segmentation step and the second step is the extraction and classification step. These methods can detect weeds and can demarcate crops from weeds. The vegetation indices can be classified into two primary categories. They are vegetation indices based on hyperspectral data/multispectral data and vegetation indices based on the visible spectrum. Ratio Vegetation Index or RVI was classified as the first indices developed. We can demarcate soil and vegetation easily. The main problem is with the variable optical properties of the soils. After RVI, NDVI has become popular. The Vegetation indices that mainly use multispectral images are Ratio Vegetation Index or RVI, NDVI, NDRE or Normalized Difference Red Edge Index and GNDVI or Green Normalized Difference Vegetation Index. Some of the RGB based vegetation indices are Normalized difference index or NDI and Excess Greenness index or ExG.

10.3 DRONES

Drone markets can be classified as heterogeneous in nature. There are several models in the drone market. They can be classified according to size and capabilities. We can call them small drones and large drones. Drones with multi-rotor configuration can maintain stability and we can easily pilot these drones. The drones with a single rotor can be the same as a helicopter. It has more efficiency. This can be expensive and piloting these drones can be hard. Drones with a fixed wing configuration can be the same as an airplane. We cannot easily pilot these drones but they can cover vast areas within less time frames. Artificial intelligence, Big data and the Internet of Things can be considered as important technologies that can determine the efficacy of drones. The drones can generate 3D maps and the costs incurred in operation will be less. We can design seed-planting patterns easily. We can use drones in nitrogen management. In the spring season, drones can be used in soil analysis, drainage analysis, tile and tillage. In the summer season, we can perform gap analysis and can do stand count. They can be used in irrigation management. We can easily observe variability in the growth patterns. We can easily monitor the crop stage and can assess the nitrogen requirements easily. In the fall season, drones can be used in the pre-harvest and post-harvest situations. We can observe stand consistency and dry down in pre-harvest situations.

Agriculture can involve the cultivation of plants, animals, fungi and other forms of life for obtaining food, medicines, fiber, biofuels and other products that can help humans in their survival. We can classify agriculture as a comprehensive business that involves many fields under one roof. Most of the products derived from agricultural practices can yield economic and social benefits. UAVs can be classified as an important alternative in remote sensing technology. We can easily perform data analysis with the data procured from the UAVs and associated platforms.

Chapter 11

Hyperspectral Remote Sensing for Crops

CONTENTS

11.1 INTRODUCTION

The narrow bands of hyperspectral images (Figure 11.1) can be important in enhancing the quality of information. We can this information to demarcate the plant stress conditions. We can ben know the green vegetation cover that exists and we can easily differentiate the variation in the greenness levels. We can also know the variations that may exist in the crop moisture. These narrow bands can be used in determining the health of the crops. The light that is being reflected can be in the range of Mid-IR, Near-IR and visible wavelengths. These wavelengths can be used in detecting insect infestations, weeds, disease and nutrient deficiencies.

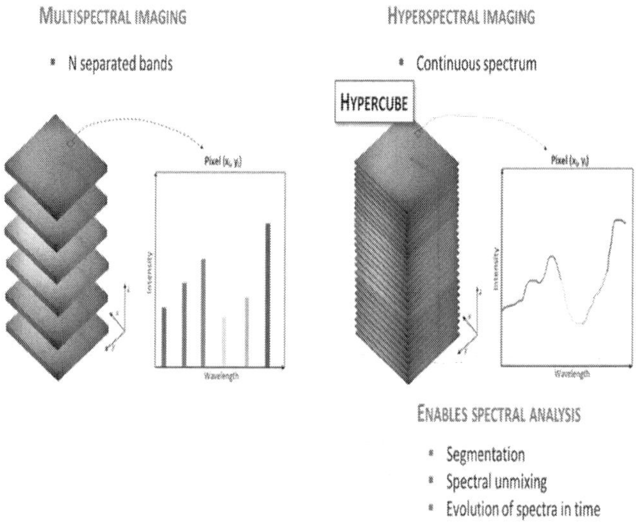

Figure 11.1: Hyperspectral images.

Source: https://www.nireos.com/hyperspectral-imaging/

The physiological stresses and diseases can alter the reflectance properties of the plants. Most of these variations can be obvious in the visible wavelengths. We can observe limited variations in the IR regions (Knipling, 1970). There are many sources from which we can get the multispectral images and hyperspectral images. We can perform quantitative analysis using remote sensing data and this can be useful in detecting the diseases and we can control them as and when needed. The hyperspectral data can provide us several reflectance values of the crops and this can in narrow bands. We can analyze these bands and there will be several bands that are to be studied. This process is complicated and will consume more time. In order to make it accessible and less complicated, there were some attempts made to limit the narrowband sets for discriminating crops. There are many methods that can be employed for this purpose. A discriminative analysis is one among many that can be used in determining the irrigation and fertilizer application rates. We can easily identify the crops that are in ecophysiological stresses (Leonardi et al., 1994). There are two important techniques that can be used in the feature reduction process in the case of hyperspectral data processing.

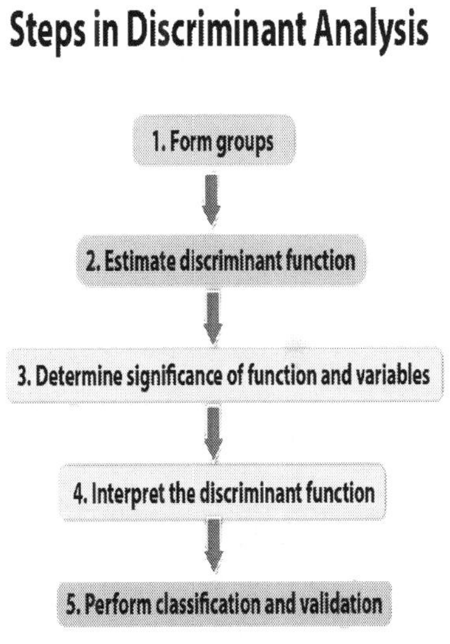

Figure 11.2: Stepwise discriminant analysis.

Source: https://data-flair.training/blogs/stat-discriminant-analysis/

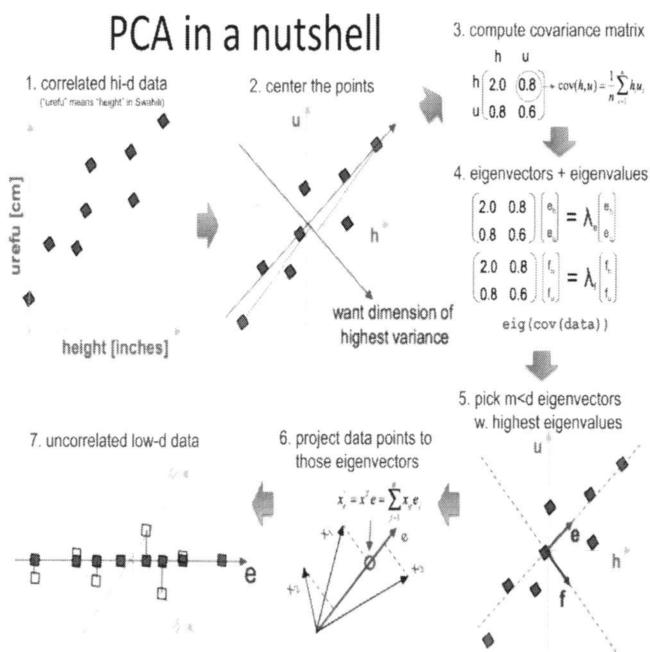

Figure 11.3: Principal Component Analysis.

Source: https://devopedia.org/principal-component-analysis

They are called as Stepwise discriminant analysis (Figure 11.2) or SDA and Principal Component Analysis (Figure 11.3) or PCA. The PCA can be used in producing uncorrelated variables or orthogonal variables. Due to this, there will be a reduction in the quantity of the variables. PCA can be used to detect the controlling variables or underlying variables or we can simply call them as factors. These factors can describe the correlation pattern in the observed variables. We can use PCA in the processing of simplifying or reducing data. This can be useful in the identification of a smaller number of factors or the minimal number of factors that can describe the variance that can be possible with a large number of variables or manifest variables. SDA can also reduce the data sets. This can be applicable to the variables that get increased with the variability in the statistical group. This will minimize variability in the groups. This technique was used to lower the number of wavelength variables. This can be called as data dimensionality.

In the case of PCA, we can expect vector variables that are new and can define the greatest variability possible in the data produced. SDA can retain original variables that can describe the demarcations between the identified groups. It is important to optimize the narrow bands. We can produce hyperspectral indices with these bands. We can easily discriminate the stresses. Some of the structural indices are NDVI, SR, SAVI or soil adjusted vegetation index, MSAVI2 or modified SAVI, OSAVI or optimized SAVI, MSR or modified SR and RDVI or Renormalized difference vegetation index. The chlorophyll related indices are MCARI or modified CARI, TCARI or transformed CARI, TVI or triangular vegetation index, SIPI or structural insensitive pigment index, MCARI1 and MCAI2. Red edge indices can be important.

11.2 HYPERSPECTRAL IMAGING

The agricultural sector is being encountered with several challenges and these are mainly due to stressors like climate change, crop diseases, environmental pollution, natural resource depletion and population growth. Precision agriculture can be considered as an important solution to enhance farming capabilities. This method of agriculture can be in the form of adaptive inputs with respective outputs and lowered environmental effects. The adaptive inputs are fertilizer and water. The ensured outputs are biomass and crop yields. It is obvious that remote sensing plays an important role in detecting crop and soil variabilities. We can even obtain critical information pertaining to the site-specific management in the fields. As far as remote sensing technology is concerned, there are two types of remote sensing. Active remote sensing and passive remote sensing are popular techniques and they are classified based on the energy requirements. The optical remote sensing technique can be classified as passive remote sensing techniques and Radar and LiDAR techniques can be classified as active remote sensing techniques. We can classify passive optical remote sensing techniques into groups based on their spectral resolutions. They are classified into multispectral and hyperspectral remote sensing groups. In the case of multispectral imaging, the spectral signals can be collected in a few bands that can be discrete. In the case of hyperspectral imaging, we can detect the spectral signals in continuous channels. The spectral bandwidths are narrow and we can say that they are below ten nanometers. With this high resolution, we can extract fine spectral features of the specified targets and in this case, it is the crop plants that we are interested in. We can use SPOT images, Sentinel 2 and Landsat images in the form of multispectral images. These images can be used to extract soil and crop attributes. The hyperspectral images must be pre-processed. This will need advanced analytical methods. These methods should be used in analyzing large amounts of data. The images can have textural, spatial and spectral features. We can extract many properties of the crops and soils with high resolution. We can use many analytical methods like deep learning in the form of CNN or convolutional neural network, machine learning in the form of RF or random forest, radiative transfer modeling or RTM with PROSAIL and PROSPECT, multi-variable regression or MLR, partial least square regression or PLSR and empirical regression with linear regression (Spath et al., 1992).

Most of the hyperspectral sensors can be on-board UAVs, airplanes and satellites. We can even install or mount these sensors on close-range platforms. We can easily obtain these images with diverse temporal and spatial resolutions. There are limited hyperspectral sensors. Some of them are TianGong-1, PROBA-CHRIS and EO-1 Hyperion. These sensors are on-board satellites. We can use EO-1 Hyperion to get hyperspectral images that can be used in agriculture. We can get data in shortwave IR, Near IR and visible regions. We can expect a spatial resolution of 30m with a spectral resolution of 10nm. This data was used in several studies until 2017 for several agricultural studies. The data obtained from this sensor was used to know the soil and crop properties, along with these applications, we can identify crop diseases. We can determine the soil features, classify the crops and even assess the residues in the crop fields. We can estimate biomass, LAI and chlorophyll. We can use this data with several classification algorithms. Some of them are adaptive coherence estimator and spectral angle mapper. These algorithms can be used in the identification of the crops. We can even know the soil organic carbon or SOC with the data obtained from Hyperion. The spectroradiometer can be used along with the images derived from Hyperion to get better results. We can compare Landsat multispectral imagery with Hyperion hyperspectral imagery to know the details pertaining to crop productivity. We can get a high rate of classification accuracy with hyperspectral imagery (Bostan et al., 2016). The spectral range in nm for Hyperion is from 357 to 2576nm and for HyMap the range will be between 440 to 2500nm. The spectral bands in Hyperion are 220 and in the case of Headwall Hyperspec, we can get 270 spectral bands in Nanometers and 324 in micrometers. The spectral resolution of Hyperion will be 10 nm and for CASI, it will be less than 3.5. Hyperion was operated by NASA, USA.

PROBA-CHRIS can be classified as another important hyperspectral sensor on-board satellite. We use this data to extract PAR or FAPAR-Fraction of absorbed photosynthetically active radiation that might be present in the crop fields. We can also get a fraction of vegetation cover or fCover with this data. We can use multi-angle PROBA-CHRIS images to know the growth stages of the wheat crop. For these observations, we need to optimize the angles for enhanced visibility. The main issue with this sensor is that we get low spatial resolution. Hyperspectral Imager or HYSI is on-board IMS-1 or Indian Microsatellite-1. This sensor can collect spectral signals within the wavelengths starting from 400 to 950nm. The expected spatial resolution is 550m. We can use this data to analyze soil salinity, soil moisture and other important agricultural features. We can use this data to classify crops. We may not use this data for precision farming and this can be due to limited data availability and low spatial resolutions.

Temporal resolution is considered an important factor along with spectral and spatial resolutions. The temporal resolution of the sensors can limit the sensor capabilities to be used in precision agriculture. This can be attributed to the long revisiting cycles. We will miss many growth stages that can help us identify the crop stress beforehand. Cloud contamination with the images can be considered as a serious issue in addition to low temporal resolution capabilities. In some cases, we get data of low quality that can limit our interpretation. There will be a low signal to noise ratio in the case of Hyperion data. This can be dominant in the shortwave IR range.

We can use hyperspectral imaging using airborne devices. These devices can extract the hyperspectral images of the crop fields and help us in the interpretation of the crop conditions. AVIRIS was one such sensor used in earlier days. AVIRIS stands for airborne visible IR imaging spectrometer. This sensor can get the spectral signals in at least 224 bands. The range will be up to SWIR starting from visible wavelengths. We can use this data to monitor agricultural activities. Some of the vegetation properties like water content, chlorophyll, LAI and yields can be studied using this data. We can analyze tillage activities and pest infestation using this data. Along with this sensor, we have AISA Eagle, Hyperspectral Mapper or HyMap and CASI or Compact airborne spectrographic imager to get the hyperspectral images. CASI images can be used to limit the management zones, weed classification, the fraction of crop cover and chlorophyll content. We can use HyMap to know the biochemical and biophysical variables. We can know the stress signals of the crops using this data. In some cases, we can estimate the biomass and nitrogen content of the crops. AVIS images can be used in knowing the characteristics of the vegetation like chlorophyll and biomass. We can investigate crop residues with Probe-1 hyperspectral images (Bannari et al., 2009). Most of the hyperspectral images used in crop studies are from airborne platforms. AVIRIS and CASI were installed or mounted on the aircraft that flew at low altitudes. We can get 5m for HyMap and 4m for CASI spatial resolution respectively. These resolutions can help us identify the soil and crop features.

In order to compensate for the losses incurred in using airborne platforms, people started using UAVs that can be purchased at minimal costs and higher capabilities. As the low weight hyperspectral sensors are available in the market, people started mounting them on the UAVs. These sensors can give us high spatial and spectral resolution. We can estimate the vegetation phenotyping variables like leaf chlorophyll and LAI with the aid of UAVs. We can perform radiative transfer modeling with this data.

There are several hyperspectral sensors that can be lightweight and can be installed on UAVs. We can use nano-Hyperspec VNIR and Headwall Microsensors for this purpose. Some of the other lightweight sensors are HySpex VNIR, PIKA II sensor and UHD185-Firefly sensors (Skauli et al., 2009). These sensors can provide us data from 100 bands. These bands can range from visible and NIR regions. These sensors weigh around 2 kg or even less. They can be easily installed on the unmanned and manned remote sensing platforms. The sensor setup, data collection and image processing are the essential phases in the deployment of UAVs. Multi-rotor type UAVs are popular in extracting information from the crop yields. Fixed-wing planes can be useful for this purpose with higher accuracy. It is important to ensure that slow flight at low flying altitude be maintained whilst sorties. This will generate a high signal to noise ratio with better spatial and spectral resolutions. Multi-rotor UAVs can permit vertical take-off and landing or VTOL, manageable or manipulative fight speeds and low flight altitude capabilities. In the case of fixed-wing UAVs, there will be minimal speed and flight altitudes that can help in landing and take-off of these UAVs.

In general, any hyperspectral imaging system needs a hyperspectral sensor, DPU or data processing unit, GPS and IMU. The weight should be less at all times so that the flight parameters can be met without any additional technical interventions. The battery endurance and payload capacity of these UAVs can be challenging. The multi-rotors UAVs can be installed with batteries with high performance and with short endurance, 20 minutes per se or even less. In some cases, the endurance can be 3 minutes also (Lucieer et al., 2016). Most of the fixed-wing UAVs may demand high fuel capacities so that longer endurance can be possible. We can use this type of UAVs for extended flight operations. These UAVs are large and massive. These UAVs can weigh about 14kg with a 5m wingspan. The main parameters that are to be considered whilst planning UAV sorties are size, weight and power along with altitude, time aloft and geographical coverage. It is essential to procure permission from ATC or Air Traffic Control prior to UAV operation.

The hyperspectral imaging can be of close range too. This can be ground-based and lab-based hyperspectral imaging. With these techniques, we can get one cm accuracy and that can be considered as super high spatial accuracy. This type of imaging technology can be used in understanding the fine-scale properties of canopy and leaves. We can easily understand the crop stress and diseases. These sensors can be installed on both static and moving platforms. We can install them on trucks, scaffolds and linear stages. We can deploy these sensors outdoors and indoors. In this case, we can use both the sun and if needed halogen lamps as light sources.

The hyperspectral images can be obtained from sensors in their raw format or we can say in digital numbers. It is important to pre-process this raw-data. The pre-processing stage can comprise spectral correction, atmospheric and radiometric corrections. With the pre-processing stage, we can get spectral information with high accuracy. There are many approaches that can be used to analyze the data. Some researchers use linear regression methods and other methods. In the pre-processing step, we can even do orthorectification along with the geometric correction. These preprocessing stages will be monitored by the data providers. In order to perform these steps, we need GCPs or ground control points and DEM or digital elevation model. These two entities can be important in geometric correction and orthorectification. In case UAVs are operated in a push broom mode, we need to see that the IMU collects data pertaining to this characteristic. We can use PARGE and ENVI software to process this data.

The radiometric correction can be used to convert the digital numbers into radiance. This can be done with the aid of calibration coefficients. These calibration parameters will be given by the manufacturers of the sensor. It is essential that we need to update this information as and when the need arises or unless specified by the company. There will be gradual degradation of the spectral machinery that was used in the sensors and this demands a continuous upgradation of the parameters needed for calibration.

Another essential aspect is the atmospheric correction. Though these UAVs are being operated at low altitudes, they suffer from various noises in the atmosphere. This comes with the scattering by carbon dioxide, water and oxygen. It is essential to go for atmospheric correction algorithms to yield spectral information of enhanced quality. We can use the SMARTS model, empirical line correction, PCI geomatics, ENVI FLAASH and MODTRAN in the atmospheric correction step.

The hyperspectral images can have many bands and these are correlated. It is important to go for dimension reduction in the pre-processing stage. We can use principal component analysis or PCA to know the principal components that describe the entire information. Other methods that can be used in the dimension reduction are singular value decomposition or SVD, minimum noise fraction or MNF along with PCA. We can also use artificial neural networks or ANNs, wavelet transforms and uniform feature design or UMD for this operation. We can use deep learning-based approach in dimensionality reduction that is based on auto-encoder. The most popular dimension reduction method is the wavelet transform. In this method, we can degrade a signal into a series. These series are scaled versions of the actual wavelet function. This will help us to know the difference between wavelets and it will be based on the information about the frequency. This can be used to get the information about the features or we can localize the features. This is called local spectral variation. This technique can be used in image classification, feature extraction and image fusion. Along with the dimensionality reduction, we need to consider band selection and band sensitivity.

These techniques can be used in minimizing the data size. This can be done by selecting certain bands that can be classified as sensitive to the physical object present on the surface. In band selection, we can use a fast volume gradient-based method. This can be called an unsupervised method. In this method, we can easily remove repetitive bands or redundant bands. This will be based on the gradient volume. There is another important method called a column subset method. In this method, we can enhance the volume of column subsets or bands. This method can be useful in the case of noisy bands, Manifold ranking based salient band selection comprises of using band vectors. There will be ranking based o bands. This will help us in measuring the band difference.

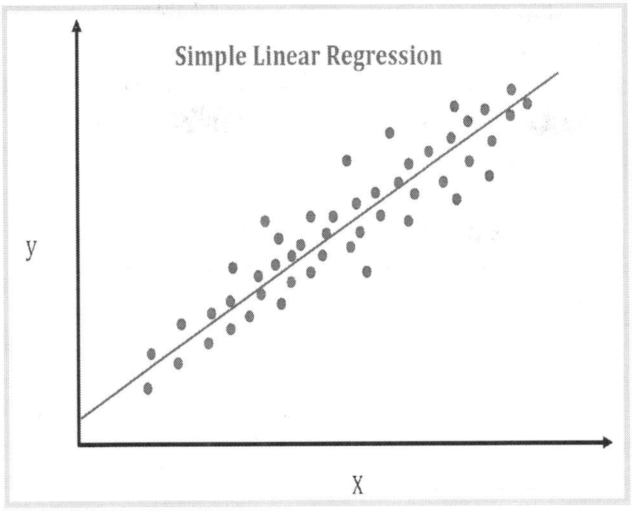

Figure 11.4: Linear regression.

Source: https://medium.com/towards-artificial-intelligence/linear-regression-basics-for-absolute-beginners-68ed9ff980ae

Linear regression (Figure 11.4) can be classified as an essential method in the process of analyzing the hyperspectral images. With this method, we can easily extract the information of the target. The vegetation indices and spectral reflectance can be employed as predictor variables. These variables can help us in framing a linear relationship. We can use spectral bands produced by the linear regressions between spectral reflectance and soil moisture data from the field stations. We can know the identified bands that can correlate with the moisture. Linear regression can be used in the estimation of both soil and crop properties. In this process, we can use index-based regressions. Regression can be based on both the quality and size of data. It is also dependent on the indices. We can use advanced regression methods like PLSR and MLR. In case of advanced regression models, we can use multiple predictor variables so that we can obtain enhanced accuracy levels.

Among several models, PLSR is a widely accepted one. We can study biochemical variables and biophysical properties of crops using PLSR based regression models. We can use the PLSR model with R and Python platforms. We can study water, grasslands and forests with this model.

This model can help us in solving the over-fitting problem and data-noise problems. We can easily tackle large predictor variables that are inter-correlated. Several attempts were made to enhance the accuracy of the PLSR model. We can add diverse predictor variables like textural variables, indices and spectral bands to increase the accuracy of the PLSR model. We can use PRESS or predicted residual error sum of squares statistics to know the latent variables. We can extract and evaluate features that can be used to select the predictor variables. With all these attempts, we can easily obtain optimal model accuracy.

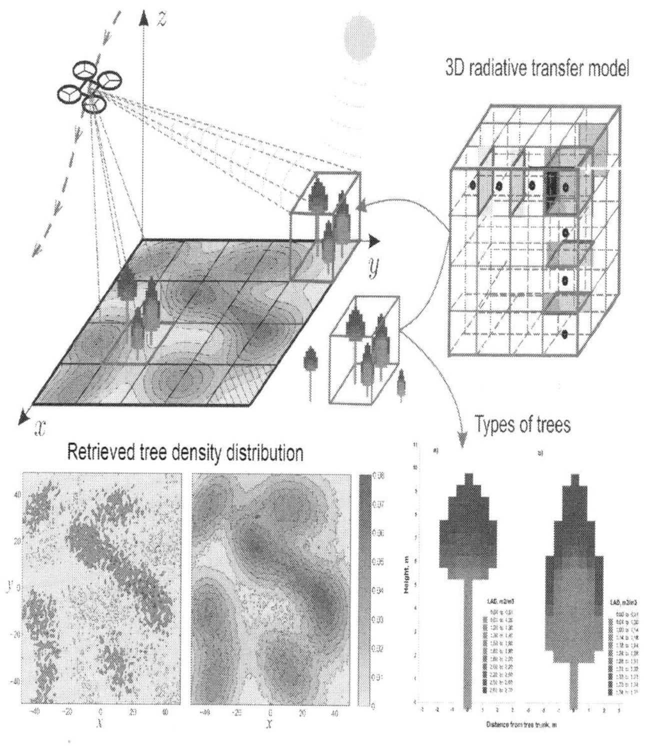

Figure 11.5: Radiative transfer modeling.

Source: https://www.mdpi.com/2072-4292/10/10/1661

Radiative transfer modeling (Figure 11.5) or RTM can be classified as a physical approach that can be used to simulate the EMR interaction with vegetation via reflection, absorption and transmission. RTMs can be useful to simulate vegetation characteristics like transmittance and leaf reflectance. These RTMs can be used in understanding the biochemical and biophysical properties in the forward mode. We can also use the inverse mode. RTMs can be used in most of the applications and we can use PROSAIL. This model can be used in the simulation of canopy reflectance with the aid of leaf properties like water content and chlorophyll, soil reflectance and structural parameters of the canopy. This can be employed in diverse agricultural environments to know soil and crop properties.

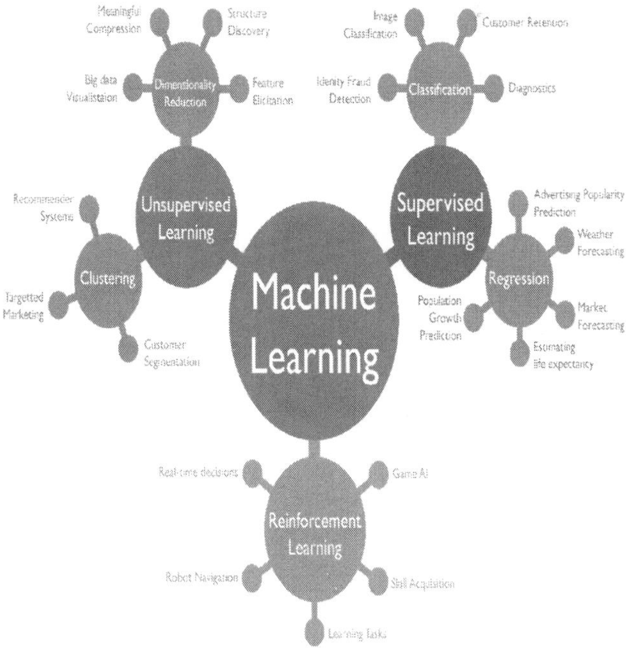

Figure 11.6: Machine learning algorithms.

Source: https://towardsdatascience.com/machine-learning-algorithms-in-laymans-terms-part-1-d0368d769a7b

Machine learning algorithms (Figure 11.6) can be used in determining crop and soil properties. We can use the Random Forest model or RF and support vector machine regression or SVM to analyze the hyperspectral images. With these algorithms, we can easily process variables. We can use machine learning tools in remote sensing technology to know the ground covers and classify them. We can use SVM as an algorithm in the classification step. We can estimate and quantify the biomass of the crop with UAV and SVM. We can easily classify crop types with enhanced accuracy. The tillage practices can be studied with the aid of SVM and KNN classifiers. Random Forests can be used in understanding diverse agricultural features. We can identify and classify weeds and crop types with RF.

Deep learning is yet another important method that can be used for this purpose. It can be called as a subset in the machine learning process. We can easily represent data as a hierarchical representation of the dataset that may be present in the model. Deep learning can be used in the classification of images. We can easily classify land cover and ground features with deep learning algorithms. This technique can be used in precision farming in order to address complicated issues. We can easily estimate the crop yields with the aid of multispectral images and CNN along with climate data. We can easily classify land cover in the agricultural fields by using multi-temporal SAR images and deep recurrent neural networks. The crop classification can be done with the multispectral images having high temporal resolution and 3D CNN (Ji et al., 2018). We can detect plant diseases with the help of images procured from smartphones and CNN. In a nutshell, we can say that both deep learning and machine learning methods can be used in processing multi-type and multi-source data within less time.

In Linear Regression, we can use one predictor variable and that can be vegetation index and reflectance. There will be a response variable like chlorophyll in this method. In the advanced regression model, we may need multiple predictor variables. It is essential to use the response variable. We need to consider parameters in the model like latent variables. In the case of radiative transfer modeling, we can use many predictor variables and parameters. In the case of machine learning, we need to use multiple predictor variables, response variables and parameters present in the model. In the case of deep learning, we can use predictor variables, the number of layers along with weights and sizes of the layers. The model complexity is low in the case of linear regression. The model complexity is medium in the case of advanced regression and machine learning. We can expect high model complexity with RTMs and deep learning.

There are many applications that can be derived from hyperspectral imaging in agriculture. We can know the biochemical properties of the plants.

We can classify the reflectance spectra in order to define the plant species and crop types. This is due to the fact that hyperspectral imaging has become prevalent in many agricultural sectors and they are associated with the field spectroradiometers and imaging spectrometers. The hyperspectral measurements comprise contiguous measurements or wavebands that are finely spaced and are numerous. We get a high rate of dimensionality and so large data volumes. Most of these instruments work at wavelengths starting from 350 to 2500nm. The bandwidths can be from 1 to 10nm. In order to obtain a training sample, it is essential to use a large sample size that can be representative of all the similar features in the image. Due to high dimensionality, the hyperspectral data will suffer from the Hughes phenomenon. The excess quantum of data can come with noise and this can be managed with minimal sample size. The imaging and non-imaging devices can yield sampling intervals of diverse nature. We cannot easily compare these two wavebands. We can aggregate the wavebands into bins of 50nm and this will be based on the band centers. The spectra can be analyzed in the lab using several analytical methods. We can use derivative analysis in this case. In this technique, we may need reflectance spectra that can be classified as smooth. We need to generate smoothing algorithms. This will preserve the local spectral features. The noise will be removed from the data. Noise can occur in different intensities and this can be present in several bandwidths. The smoothing process for a particular signal exists in the form of algorithms. We can use mean filters and median filters.

The plant stress can be seen in the canopy and it carries several symptoms. The temperature and color of the leaves will be changed in water-stressed conditions. The plant water stress can be due to a combination of environmental conditions or due to a single parameter. The sap flow and leaf water potential can give users direct information pertaining to water conditions. In this case, we may need destructive sampling or plant contact. This is not feasible. The non-destructive and non-contact sensing techniques can monitor plants and even we can prefer automated sensing capabilities. Plants provide a dynamic response. The water content in the leaves can be measured with Mid IR and NIR spectral regions.

The data mining techniques can be important in vegetation studies. They can be used in the feature extraction. The information content includes information entropy, band variance and selection that is based on theoretical knowledge. The projection-based methods include principal component analysis or PCA and independent component analysis or ICA. The diverse measures include distance-based measures and others. The similarity measures include spectral derivative analysis, correlation coefficient ,etc. The wavelet decomposition method can be classified as an important technique. There are several methods that can be used in the classification of vegetation.

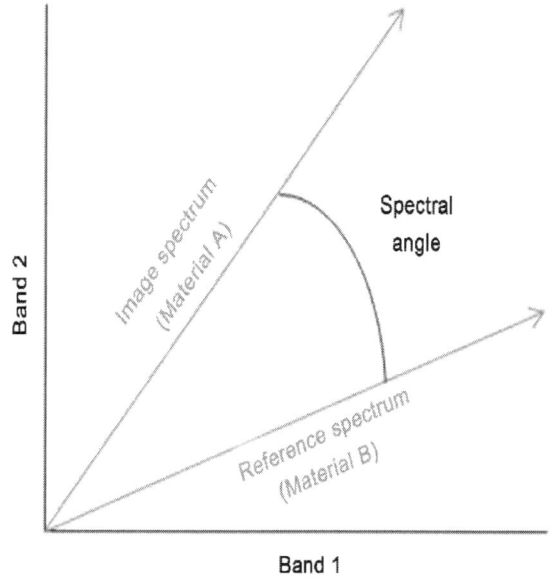

Figure 11.7: Spectral angle mapping.

Source: https://towardsdatascience.com/machine-learning-algorithms-in-lay-mans-terms-part-1-d0368d769a7b

They are spectral angle mapping (Figure 11.7) or SAM, Maximum likelihood classification (Figure 11.8)or MLC, ANN or artificial neural network 9Figure 11.9), SVM or support vector machines (Figure 11.10) and SMT or spectral matching technique. The spectral mapping technique can be used in the spectral analysis and it may need a spectral library. We can generate hyperspectral vegetation indices. There are some properties that can be classified as unique. We can easily eliminate the redundant bands and that means the removal of highly correlation bands. We can get physically meaningful HVIs. It can be photochemical reflective index or PRI and this can be used in estimating light use efficiency. We can enhance the quality of the broadband indices. We can reduce broadband saturations.

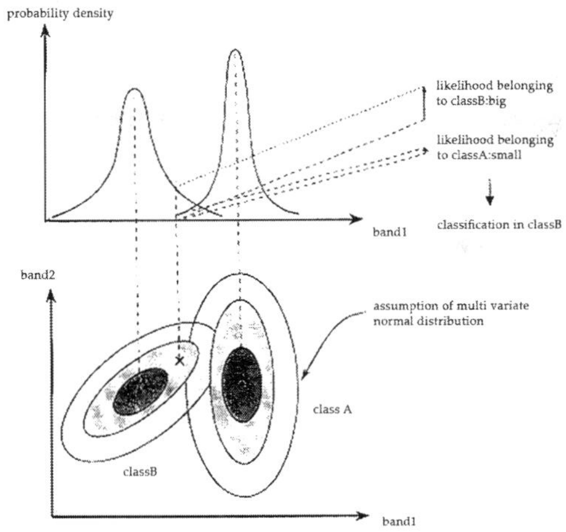

Figure 11.8: Maximum likelihood classification.

Source: http://sar.kangwon.ac.kr/etc/rs_note/rsnote/cp11/cp11-7.htm

We can achieve a high rate of sensitivity. We can get novel indices that might not be sampled into broadbands. The water-based indices can be classified as novel . There are multi-linear indices that can use more than two bands. The two band vegetation indices can comprise unique indices that can be used in the vegetation mapping.

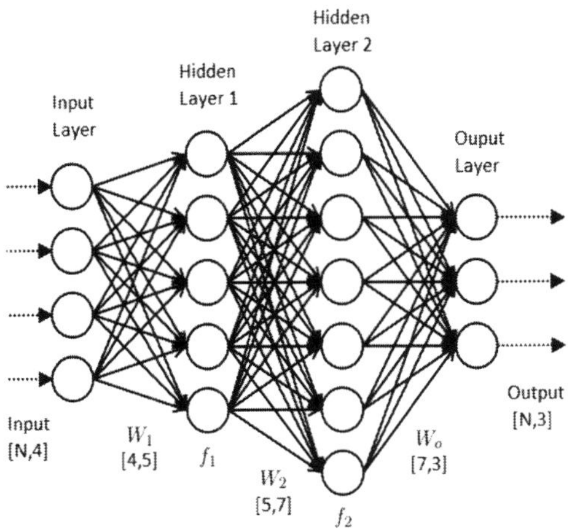

Figure 11.9: Artificial neural network.

Source: http://sar.kangwon.ac.kr/etc/rs_note/rsnote/cp11/cp11-7.htm

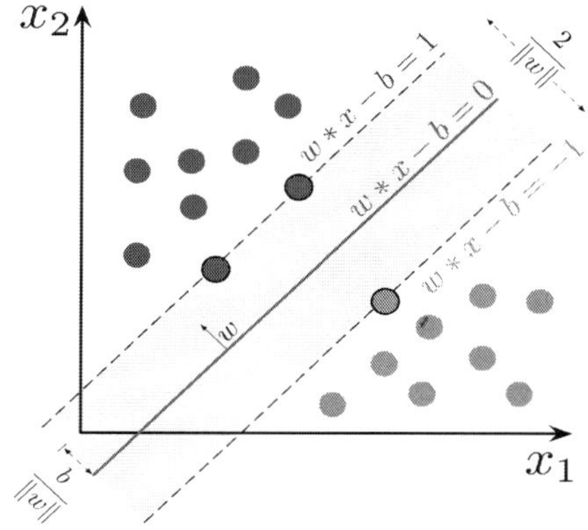

Figure 11.10: Support vector machines.

Source: https://en.wikipedia.org/wiki/Support_vector_machine

Chapter

12

Seasonal Variability Studies

CONTENTS

12.1 INTRODUCTION

The crops can be sensitive to the perils caused by weather events and this is evident in the growing seasons. The climate vulnerabilities can be classified as critical problems in the agricultural sector. The farmers face many problems with the in-season crop variabilities as most of these effects consume a lot of resources. The crops can be classified as a business with perils and this can be observed in the poor segments of society. Agriculture can be defined as the production of crops for food and other services. We can enhance the crop yields with continued research. We can produce enhanced plant varieties and with them, we can also produce agrochemicals that can enhance the production rates. Most of the producers will rely on the science tools to identify and manage the quality of the soils. These tools will be based on location and so we can manage crop parameters according to their location only. All tools might not work in all locations. The farmers are sensitive to natural forces. This can be called as seasonal climate variability. Any slightest shift from normal climatic conditions can surface production risks like diseases and pests. High rainfall rates, high humidity and high temperatures can enhance the occurrence of fungal diseases. These conditions can enhance the growth conditions for pests and insects and hence there will be horizontal transfer and spread of diseases.

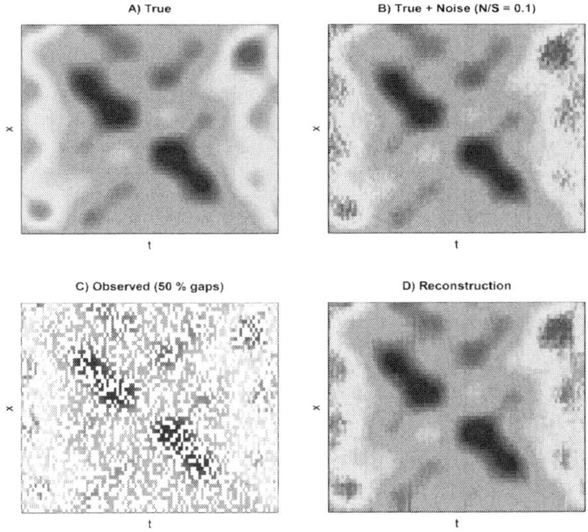

Figure 12.1: Empirical Orthogonal Function Analysis.

Source: http://menugget.blogspot.com/2012/10/dineof-data-interpolating-empirical.html

Empirical Orthogonal Function Analysis (Figure 12.1) or EOF can be utilized in the delineation of rainfall. We can observe the interrelation between time and space variables. Based on these variables, we can know the structure of other variables too. We can easily use EOF in identifying the trends of rainfall occurrence. This analysis may not be based on a hypothesis framed whilst identifying the data distribution. We can also use principal component analysis also. This can be used to integrate the space and time variables. We can use these methods to drastically lower the data dimension obtained in rainfall data sets. We can easily filter out lesser factors that can easily help us in knowing the variance percentage in the datasets. Most of the redundancies observed in the rainfall datasets will be eliminated. We can also filter the atmospheric noise and can compress data for further analysis.

In most of the models where the principal component is considered, we can notice a linear combination of the variables that might be present in the datasets. It is assumed in that way mostly. If data imbibe low variance, we can assume that the components in question are having noise. If data is exhibiting a high level of variance, we can say that this data can be used to represent dynamics in a better way. In the case of factor analysis, we can generate uncorrelated components that can be represented in the form of orthogonal lines. They can represent the best fit of the dataset under observation. We can use a standardized precipitation index or SPI in the analysis of datasets. These datasets can be obtained from the observations of the previous years.

This index can be used in assessing the rainfall shortage and drought incidence. We can understand the severity of the rough weather events. We can use the Mann-Kendall test to understand the seasonal rainfall analysis. We can also use Sen's slope estimator to know the magnitude of the rainfall data. We can conduct a quantitative analysis of the crop yields and rainfall. This can be done by regression and correlation techniques. We can use multivariate linear regression analysis, R and Pearson's correlation coefficient to frame a linear relationship between the crop yields and rainfall. We can use Pearson's correlation coefficient to evaluate the strength of the linear relationship between rainfall variability and crop variability. We can give a score of 0 if the variables under study are independent and we can assign a score of -1 and +1 in case we observe indirect and direct dependence is any between the variables. We can use the normality test to know if the seasonal rainfall followed normal distribution or not. Every year, we can observe weather-induced shocks and crop yields and this can be derived as a form of risk that can affect both supply and demand. Entire global markets are affected by weather shocks. Both consumers and producers will be dearly affected by the aftereffects of these perils caused due to weather events and climate modifications. Due to global warming, there will be weather variability and this is apparent with the frequent occurrence of heatwaves and their intensity. It is important to know the yield variabilities in this situation so as to reach a meaningful conclusion. We can use empirical models in this case. We can use empirical and process-based approaches to understand the variables. With this data, we can forecast the due impacts of pests, weeds and diseases. In order to know better, it became essential for us to generate global gridded crop models. These models can provide us a pseudo-natural representation of phenology and plant development. We can even understand maturity, anthesis and timing of planting, etc. with these models in place. Studies on both spatial and temporal scales can help us know the seasonal window so that we can forecast the weather conditions that might affect the crops. There are several factors that are not linked with weather events we need to consider. Successful simulations may need local information about the parameters that can drive crop growth patterns. For this to be valid, we need to collect information from the archives of that place. We need to consider phenological heat units or PHU and cumulative heat measurements to forecast crop maturity. The PHU along with other variables can be injected into the global gridded crop model. We can use PHU data along with the geographic variations to select appropriate crop types. In the case of crops grown in cold weather/climate, we can use minimal heat sums. This data can help us know the crop maturity levels. This can be much complicated with the crops grown in the warmer climates.

We can compare the models and if essential we can improve or models according to our needs. Most of the simulation models are being evaluated. They are based on variance. Most of the populations in the world are based on food products that are produced in the native crop fields or imported. This situation is prevalent in developing countries where agriculture is considered as a backbone of a nation. In these countries, the farmers are dependent on off-farm and on-farm agricultural labor. There are many factors that are not linked with climate and they affect crop production frequently. The climate shift has already changed the course of agriculture and most of the consequences are evident in most areas. In any disastrous climate event, we can say that most of the African countries will be vulnerable and this is due to reduced imports to them. Most of the SSA or Sub-Saharan countries are dependent on agriculture that is heavily dependent on weather. Most of the investments were poured into the policy-making and execution in the direction of climate mitigation resulting in resilient agriculture. The variations in the extremes can pose a direct threat for agriculture and this can be considered as an important issue. Due to extremely high temperatures in summer, most of the food production got reduced and due to this several people were perished. The availability of fruits and cereals was diminished in Europe and so the maize crops. Limited information exists about the extreme events and this is true as advanced recording devices were of low grade in the olden days. There were attempts made to project the frequency and severity of climate events and their impact on crops. Some simulation models also explained the negative and disastrous effects of climate variability on crops. The extreme daily temperatures can damage the crops within no time. Statistical models were used to understand the inter-seasonal variability in precipitation and temperature on the crops. It is important to determine the possible effects of climate and weather events on crop production. For this, we need to know the cross-sectional spatial and temporal attributes. It is essential to frame linear mixed models in the case of sorghum, rice and maize crops. We can use longitudinal data to appropriate linear mixed models.

The spatiotemporal rainfall variability can be called a SRTV model. This model can be used in research and development in agricultural research.

12.2 CLIMATE RESILIENT AGRICULTURE

The climate linked hazards are often considered disastrous and the consequences can be harmful to poor sections. It is essential to promote the concept of climate-resilient agriculture to cope with the problems associated with climate borne risks. This will ensure food and income security in the farming communities. It is projected that climate change can exacerbate the perils in the coming decades. There might be intense floods and droughts. We can expect drought and this will be a starting point to dwindle the economies unless addressed in a correct and organized way. Most of the hazard events can impose a direct threat to agricultural production. The main areas that are subjected to damage are livestock keeping and crop farming. It is estimated that there will be conflicts, migration and degraded ecosystems due to problems derived from the climate change scenarios. All these effects can lead to severe economic losses and health issues. It should be understood that the pastoralists and farmers need to upgrade agricultural practices so that they can tackle climate borne perils. It can be known that some of the agricultural policies be modified so as to bring about change in the current farming operations. Most of the systems or communities that are exposed or vulnerable should be able to forecast the risks and they should respond in a timely manner with foolproof adaptive measures, then we can call such communities or systems resilient for climate initiated perils. We can reduce vulnerability by altering exposure. This can be done by assessing the impacts and then we can even map the hazards accordingly. We can plan land use land cover and conduct surveys. We need to protect watersheds and this will assist us in establishing retention zones that can safeguard the watersheds. We can restructure agriculture and help the exposed populations. We can alter the cropping patterns. We can reduce sensitivity. This can be done by adopting or developing animal varieties and select appropriate crops that are resistant or tolerant to climate risks. We can enhance the quality of the drainage systems and irrigation systems. We should improve water management and soil nutrition practices. We need to diversify the cropping systems. We need to set certain standards that can help us in the adaptations whilst disasters. We can also enhance adaptive capacity. This can be done by developing action plans and adaptive strategies. We can develop capacities that can implement plans and might integrate with many entry points. We can diversify income sources. We can enhance infrastructure systems and enhance water frameworks. It is essential to establish crop insurance schemes and disaster-resilient schemes. We need to promote capacity building and technology transfer. We can define vulnerability as a degree to which any system or systems is sensitive to the disturbing effects of climate change. (Figure 12.2)

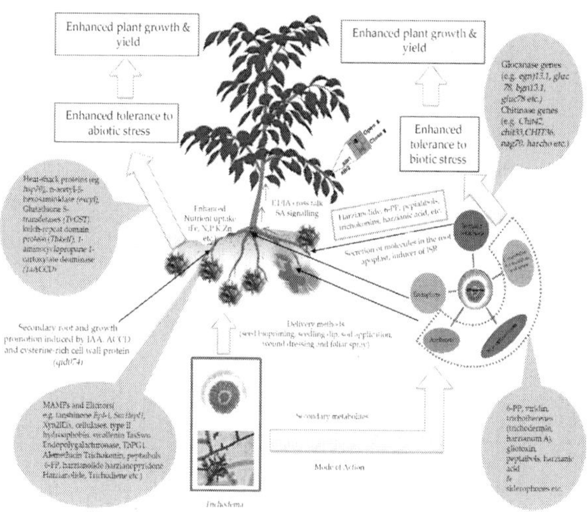

Figure 12.2: Climate resilient agriculture.

Source: https://www.semanticscholar.org/paper/Trichoderma-for-climate-resilient-agriculture-Kashyap-Rai/8dff215c99d81aa0b03c79ad12d0155133c3da24/figure/0

The concept of resilience can help us understand the system organization in multiple states of stability and this will be based on the situations faced by the individuals. Any rancher will prefer to grow grasses so that his economy will be justified and labor costs will be recovered. In the case of small and marginal farmers, shrubs can be useful as they produce certain amounts of yields that can be used in the current situations without any problems. We need to observe and record the thresholds of a system that can be classified as boundaries. Once the parameters shift from these boundaries either vertically or horizontally, we can classify it as the transition. Due to this, there will be transferred system change. In some cases, we can observe changes in the policy thresholds, social and ecological thresholds, then we can call it is a transformational change. Whilst decision-makers come up with many plans that can help us combat the problems with climate change, in some cases, these plans can become untenable and obsolete. This demands alternative strategies. In the agricultural communities, the resilience may not be restricted to crop production itself but it extends to several sections of society. It will extend to several commodities and this will demand change. It is essential to use commercial opportunities and social networks to extend the practices over several domains. This can be called permanent and temporary migration. Most of the resilience strategies can explain the tendencies of the system to react or respond to climate change. This tendency can be based on enhanced adaptive capacity, lowered sensitivity, resources and information.

Most of these entities are distributed randomly within a population. We can say that individuals have diverse adaptive capacities and some of these aspects might need adaptive capabilities. In some cases, one aspect can compensate other automatically without our intervention.

The climate-resilient agriculture was construed as climate -agriculture. This has a tendency to integrate three sectors that are assumed to be a part of sustainable development. They are economic, social and environmental dimensions. These domains or dimensions can address climate challenges and food security. We need to enhanced agricultural productivity with sustainability intact. This will help us enhance the farm incomes, food security and allied benefits. It is essential to start the adapting process and the resilience must be constructed at multiple levels. We need to reduce greenhouse gas emissions that might be generated whilst agriculture practices. We need to enhance the carbon sequestration process. Food security can be considered as the main objective in climate-smart agriculture. This can be explained in several aspects. The production of crops can define food availability or supplies and hence food chains will be maintained. The accessibility of food crops can be considered as important. Then comes the accessibility of food products. This will determine the prices in the market and so demand and supply will be in place and it can be considered as a determining factor. Stability can be considered as an important dimension in resilience. This will be based on value chains and production systems. Then comes utilization as it is essential to bring about nutrient quality. This can be based on processing stages.

It is assumed that even if there is no climate change, there will be a shortage of food. The climate-smart agriculture will ensure adaptation and resilience to climate change, lowered GHG emissions and enhanced crop production. We can enhance soil organic matter with certain agricultural practices. We can go for alternate drying and wetting patterns in the case of swamp rice. We can ensure zero grazings of cattle and goats. The main objective of climate-resilient agriculture is to enhance food security and improve value chains in agriculture. In the case of climate-smart agriculture, we can expect food security and enhanced resilience. We can mitigate greenhouse gas emissions. In the case of sustainable agriculture, we can maintain natural resources without compromising the present needs and conserving them for the future. The sustainable livelihoods can focus on maintaining household strategies. In sustainable intensification, there will be judicious utilization of the resources along with enhanced production. In the case of climate-resilient agriculture, we can focus on climate change impacts. The geographical scale can be from regional to local scales. This aspect is both incremental and transformative. We can use this for both resilience and adaptation. The climate-resilient agriculture is in vogue now. In the case of climate-smart agriculture, we can focus on climate change and the geographical scale will be from global to local scale. This can be incremental and transformative in nature. It can deal with resilience, adaptation and mitigation. It is still in vogue. Sustainable agriculture can focus on the environment. The geographical scale covered will be from global to local scale. This can be incremental alone.

We can construct resilience with this. It is outdated and stopped in the 90s but still remains relevant in current scenarios. Sustainable livelihoods mainly focus on social wellbeing. The geographical scale will be limited to the community and households. It can be both transformative and incremental. In this case, we can cover vulnerability aspects. It is classified as outdated and it can be relevant. Sustainable intensification focus on productivity. The geographical scale will be from landscape to local scale. This can be incremental. The main response that is expected to cover can be an adaptation. It is still in vogue. There are several external factors that can control civil insecurities, financial and globalization scenarios that affect the agricultural sector as a whole (Brock, 2012). In the pre-production stage, we need to establish a favorable environment that can be used in conducting the climate change policies related to agriculture. It is essential to manage climate-proof systems even in the agricultural institutions and policies. We need to design the latest policies that can drastically lower the risks in the production systems. This can be done by consultative processes. We need to construct new institutions so that the concepts like adaptations and resilience can be adopted easily within the farming sectors. We need to review the technologies in vogue and then we need to see if any gaps exist. It is essential to fill the gaps as soon as possible via research and development. In the case of the financial sector, we need to analyze the financial implications that are due to climate change scenarios. The state and country budgets must be modified according to the needs. Most of the tax revenues will be reduced. It is essential to provide financial services to the farmers in order to mitigate climate change problems. We need to diversify and localize the financial instruments so as to enhance the farmer's behavior to yield resilience.

12.3 INFORMATION TECHNOLOGY

The IT sector can provide us information about the causes that can generate climate change. It also helps us in constructing resilience. The risks and impacts of climate change can be easily shared across communities. We can both learn and monitor the impacts of climate change on crops and other natural resources. We can encourage consistency of the climate-resilient projects in a long-term. It is important to enhance the social networks so as to spread the concept of gender equality and cohesion and this will attend to resilience. The technology management sector can identify the agricultural assets and risks from the projected climate hazards and native hazards that might be induced by climate change effects. We can identify novel approaches to manage natural resources and agricultural resources and make them resilient to after effects of climate-induced problems. In the production stage, we need to develop climate proofed security strategies to safeguard the agricultural sectors. We need to strengthen the institutions that offer services pertaining to both adaptation and resilience under one roof. We need to suggest safety nets financially to the farmers in case of perils linked with the climate change stressors and shocks. We need to review the water policy and its pricing keeping in view the hardships that farmers encounter. It is important to give weather forecasts and other relevant information pertaining to the adaptation strategies to the farmers. Most of the information products must be based on native and local knowledge obtained and it must be based on climate variability. We need to enhance the ecosystem's health and hence the buffer capacity will be upgraded as per the need. We need to develop natural resources effectively. It is important to lower the risks caused by climate change with the aid of technical products.

In post-harvest situations, it is important to develop storage and distribution capabilities. It is important to encourage farmers so that the value chains will be improved and so a reduction in the spread of risk. It is essential to install certain financial instruments to lower the risks whilst post-harvest situations. We need to use certain extension methods so that the farming communities will be well informed in advance. This will also enhance the value-chain in the post-harvest conditions. We need to develop an infrastructure that can be climate-proof and can be manageable. We need to adjust the technologies assisted with the climate facilities. We need to develop certain policy instruments that are needed to improve availability, quality and accessibility to the foods. It should promote policies that meet climate-resilient efforts. It is also important to include financial contingencies in case of emergency situations like harvest failure and famine. It is important to monitor demands. It is essential to forecast the farm products and this should be in response to the risks associated with the climate change scenarios. It is better to drastically minimize the wastage of the foods. We need to establish contingency systems in case of harvest failure.

We need to develop policies that can enhance both the availability and quality of the food products. We need to encourage policies that can entangle food-choices with climate resilience. We need to encourage crop diversification as and when needed. We need to produce resistant varieties that can withstand the temperature extremes. We can use improved varieties or hybrids so as to get enhanced crop productivity. These varieties should withstand environmental stresses and must have nutritional values. We can develop drought-resilient crops in areas where rainfall is scanty. Drip irrigation can be classified as an important adaptation strategy. It can be described as a constant supply of water to the crops. In this system, we can supply enough moisture required by the plants and hence the rooting zones will be more focused. This can reduce the wastage of water and enhance water utility. Livestock disease management can be important. Livestock diseases can devastate the production system economics. This can also affect human health, food insecurity and loss in productivity. We need to manage livestock diseases in order to obtain the required results even if climate extremes are in place. This can be done with vector control techniques, diagnostic tools, vaccines and antibiotic use. Agro-forestry can be important.

Rainwater harvesting can be important in the management of watersheds. We can use rainfall as a better source to hold watersheds in place. The rainwater must be collected via diverse approaches prior to its contamination by anthropogenic or other factors. Most of the water losses can be due to runoff, drainage, evaporation and transpiration. It is obvious that in the limited availability of surface water, we need to opt for groundwater. This groundwater can be recharged with rainfall events and the artificial recharge can be used in the case of sparse rainfall events.

Flood hazard mapping can be used in order to sustain the climate extremes. Flooding can be a sign of extremely limited management of water resources and it should be addressed appropriately. Climate change can bring unwanted floods in several places leading to disastrous consequences. We need to identify the flood zones and then map them so as to prevent or reduce damage to the communities. Flood warning systems can be used in detecting the flood events in advance and can buy us some time to plan accordingly. This system is not merely a warning board but it saves many lives. It is important to use household water treatment installations and safe storage houses for water. This can improve the quality of water at home and so we can avoid the unnecessary addition of microbes into water. We can even reduce the incidence of many diseases.

Conservation agriculture can be based on three important principles in order to intensify sustainable agriculture across several nations. Some of the fundamental practices are crop rotation, mulching with the residues obtained from the crop harvest and minimal disturbance to soil or we can say limited tillage or no tillage. In some places, it is called zero-tillage.

There is a need for smart practices and adaptative technologies that are to be maintained in climate-resilient agriculture (Mondal., 2020). The goal of conservation agriculture is to provide all the benefits to the farmers in the form of enhanced soil fertility, lowered yields and reduced input demands. In case of reduced rainfall, we can use planting basins along with mulching. This will increase moisture conservation and water infiltration. If the rainfall is high, we can keep the crop residues so that the rate of erosion will be reduced drastically. It is essential to practice conservation agriculture in different modes and it should be a part of the agricultural development policy. The identification of conservation agriculture itself can be based on the adoption patterns of the latest technology. There will be sources of endogeneity and this will cloud our judgmental capabilities. The climate change mitigation measures might need several adaptation approaches all along with the crop production systems. This can include minor changes in the crop varieties that might have surfaced due to inefficient decisions. The farmers can alter the livelihoods and this will be based on pastoralism. There are many animals that can be integrated with the farming systems and this will lower excess dependency on crop production.

Most of the farmers might need to adopt the latest farming practices to enhance crop production rates. This can be possible with the fertile soils and enhanced biodiversity. With improved soil health, the farmers can enhance the water holding capacity of the soils and so the infiltration capacities. This will improve the speed of percolation in the soils. When the soils are improved, then the uneven rainfall events that are to be deposited on the soils will easily get percolated and so we can get benefitted from climate change. The agroecological practices can improve soil fertility and health and hence productivity. We need to add compost and manure whenever required and this will bring several advantages to the soils. The essential nutrients for the soils will be added with this improved soil structure. We can use green manures, crop rotation and mulching practices to bring about a defined change in the soil structure and fertility. System resilience can enhance biological diversity. These improved agricultural practices can increase biodiversity and this will mimic the ecological process and so the risks will be lowered due to climate shocks. Research in this sector suggests that those producers/farmers who followed diversity faced minimal damage that was caused due to climate change-induced perils. The conventional farmers who followed the use of monocultures encountered many losses. In most of the cropping systems, crop diversity can be enhanced by using diverse crop varieties at a given place. This can be done by adding animals and vegetation into the system. The farmers can grow diverse plant varieties to compensate for the losses due to climate shocks. Soil health can enhance the biodiversity of the soil organisms and can enhance the soil nutrients that might result in decreased incidence of disease. It is essential to maintain diverse agroecosystems so as to adapt to the pests.

The adaptation to climate change may need excellent water harvesting and water management practices. There are several traditional techniques that are in vogue that can provide us a solution to harvest rainwater. This information must be shared among the farming communities so that the advantages of these techniques can be available to them. Climate resilience can be done with five important elements. We need to construct or frame several international policies in this matter. We need to implement a research agenda so as to bring about the climate-resilient ecosystem if possible. We might need to stop agriculture that might destruct environmental resources and high emissions agriculture should be eradicated. We need to a reduced vulnerability against climate risks. We need to procure investments so that there will be ecological agriculture instead of climate-sensitive agriculture.

Ecological agriculture can bring about resilience with enhanced adaptive capacity. This can be possible by sustaining soil fertility and quality. The soil organisms play a vital role in enhancing the quality of the soils (Nicola et al., 2019). We need to encourage communal water conservation and so the catchment systems. It is essential to go for small scale agro-forestry and this will entail degraded lands. The national frameworks on the agricultural policy should be rejuvenated and amended so as to accommodate the climate resilience agriculture and adaptations. We need to conduct several in-depth assessments so as to identify the barriers. The coherence in the agricultural policies should be ensured and maintained at all times. The subsidy priorities should be revised according to the projected scenarios. The environmental impacts should be kept in mind while drafting agricultural policies.

The climate services were installed in several areas so as to manage the impacts of climate change effects (Attri, 2020). In the Mediterranean regions, attempts to mimic exact climatic conditions were attempted (Baytorun et al., 2018). Ecological farming should be encouraged so that we can determine the subsidy proprieties and can focus on the sensitive sectors. We need to entail agroecological practices as and when required and this will enhance our resilience towards climatic variations. It is always a good practice to maintain agrobiodiversity as most of the weather patterns and climatic effects might not penetrate into the diverse ecosystems. The use of monocultures must be avoided and decent efforts pertaining to the plant breeding experiments should be done. We need to focus on adaptive capacity against vulnerability.

We need to strengthen the small-scale and marginal farmers and encourage them in agricultural decision-making. It is essential to share the information with all the farming communities. These initiatives should be participative and must invite farmers from all sectors. As a group, the problems associated with the latest additions in their farming operations should be discussed and addressed at a time so that they can take home the essential ideas that can suit them. The meteorological stations at national and regional levels should be fed with funds and their services should be used in a timely manner. Most of the efforts must be focused on pre-disaster management and later focus can be shifted to post-disaster management. We need to discourage the use of chemical pesticides by reducing the incentives and loans to those who encourage their use. We need to enhance crop productivity and this will ensure nutritional security and food security. It is important to frame a comprehensive strategy to understand the multitudes of climate change that might affect crop production rates. We need to know the impacts of climate change that were least studied earlier and we need to quantify both intensity and production of the crops. The field workability should be ensured with the work calendar. We need to record the farmer's response against climate shocks under stress. We need to understand the trend, variability and predictability of the rainfall events. We need to use the observations obtained from the local stations so as to ensure the credibility of the data procured. We need to adopt tailor -made phrases and words to spread the work about the climate scenarios.

It is obvious that we need to know the rainfall distribution patterns and crop production rates whilst framing models. In the case of high humid areas, we need to correlate rainfall data with the crop production rates. We need to use variation analysis according to the seasons. There are some uncertainties that can cloud our observations about climate change. This can be due to aerosol emissions and greenhouse gases. The global climate sensitivity can also put in into jeopardy whilst modeling process. The climate associated scenarios are coherent and we need to project the required data according to the region of interest. We need to opt for incremental scenarios. We need to use arbitrary amounts of extrapolation of the climate events and it should allow the natural climatic events to supersede our predictions. We can apply these inputs in order to explain the sensitivities in detail. We need to use analog scenarios in the development of climate scenarios. These scenarios are spatial and temporal in nature. The temporal analogs can use past climate data in order to predict and forecast future climatic scenarios. The spatial analogs can be based on the climate data obtained in particular regions and we can know the climate of the future for that region with high confidence. Climate model out-based scenarios are popular. These scenarios are based on the experiments called general circulation model experiments.

REFERENCES

1. Abbaspour-Gilandeh, Y., Fazeli, M., Roshanianfard, A., Hernández-Hernández, J. L., Penna, A. F., & Herrera-Miranda, I. (2020, 05). Effect of Different Working and Tool Parameters on Performance of Several Types of Cultivators. Agriculture, 10(5), 145. doi:10.3390/agriculture10050145
2. Adiku, S., & Stone, R. (1995, 12). Using the Southern Oscillation Index for improving rainfall prediction and agricultural water management in Ghana. Agricultural Water Management, 29(1), 85-100. doi:10.1016/0378-3774(95)01181-1
3. Al-Mulla, M., & Sepulveda, F. (2014, 05). Novel Pseudo-Wavelet Function for MMG Signal Extraction during Dynamic Fatiguing Contractions. Sensors, 14(6), 9489-9504. doi:10.3390/s140609489
4. Alchanatis, V., Schmilovitch, Z., & Meron, M. (2005, 02). In-Field Assessment of Single Leaf Nitrogen Status by Spectral Reflectance Measurements. Precision Agriculture, 6(1), 25-39. doi:10.1007/s11119-005-0682-7
5. Andrews, L. C. (2019, 01). Atmospheric Structure. Field Guide to Atmospheric Optics, Second Edition. doi:10.1117/3.2318080.ch1
6. Angus, J. F., Stapper, M., & Donnelly, J. R. (1993). Simulation models for strategic and tactical management of crops and pastures. Journal of Agricultural Meteorology, 48(5), 775-778. doi:10.2480/agrmet.48.775

7. Anonymous. (2020, 09). Useful advancement of the already available FAOSTAT estimations. doi:10.5194/essd-2020-202-rc3
8. Asner, G. P., Scurlock, J. M., & Hicke, J. A. (2003, 04). Global synthesis of leaf area index observations: Implications for ecological and remote sensing studies. Global Ecology and Biogeography, 12(3), 191-205. doi:10.1046/j.1466-822x.2003.00026.x
9. Atkins, P. J. (1988, 12). Book Review: Land Transformation in Agriculture. Outlook on Agriculture, 17(4), 187-187. doi:10.1177/003072708801700415
10. Bagnall, R. (1977). Resistance To The Aphid-Borne Viruses In The Potato. Aphids As Virus Vectors, 501-526. doi:10.1016/b978-0-12-327550-9.50026-4
11. Bannari, A., & Staenz, K. (2009, 10). Soil Backgrounds Impact Analysis on Chlorophyll Indices Using Field, Airborne and Satellite Hyperspectral Data. Geoscience and Remote Sensing. doi:10.5772/8294
12. Berliner, L. M. (1996). Hierarchical Bayesian Time Series Models. Maximum Entropy and Bayesian Methods, 15-22. doi:10.1007/978-94-011-5430-7_3
13. Bode, A., Anadón, R., Morán, X., Nogueira, E., Teira, E., & Varela, M. (2011, 08). Decadal variability in chlorophyll and primary ¬production off NW Spain. Climate Research, 48(2), 293-305. doi:10.3354/cr00935
14. Boincean, B., & Dent, D. (2019). Carbon Sequestration and Climate Change. Farming the Black Earth, 53-87. doi:10.1007/978-3-030-22533-9_4
15. Bostan, S., Ortak, M. A., Tuna, C., Akoguz, A., Sertel, E., & Ustundag, B. B. (2016, 07). Comparison of classification accuracy of co-located hyperspectral & multispectral images for agricultural purposes. 2016 Fifth International Conference on Agro-Geoinformatics (Agro-Geoinformatics). doi:10.1109/agro-geoinformatics.2016.7577671
16. Brönnimann, S. (2002). Picturing climate change. Climate Research, 22, 87-95. doi:10.3354/cr022087
17. Brock, C., Derkzen, P., Fritz, J., Neff, A. S., Florin, J., Hurter, U., & Wahl, V. (2019, 12). Evolving Agriculture and Food – Opening up Biodynamic Research. Open Agriculture, 4(1), 803-804. doi:10.1515/opag-2019-0079

18. Bruijn, F. J. (2015, 07). The Quest for Biological Nitrogen Fixation in Cereals: A Perspective and Prospective. Biological Nitrogen Fixation, 1087-1101. doi:10.1002/9781119053095.ch108
19. Brusewitz, G. H., & Stephens, C. L. (1979). Wheat Protein Variation Between Loads. Transactions of the ASAE, 22(6), 1431-1434. doi:10.13031/2013.35225
20. Bussay, A., Velde, M. V., Fumagalli, D., & Seguini, L. (2015, 12). Improving operational maize yield forecasting in Hungary. Agricultural Systems, 141, 94-106. doi:10.1016/j.agsy.2015.10.001
21. Cammarano, D., Fitzgerald, G., Casa, R., & Basso, B. (2014, 03). Assessing the Robustness of Vegetation Indices to Estimate Wheat N in Mediterranean Environments. Remote Sensing, 6(4), 2827-2844. doi:10.3390/rs6042827
22. Campbell, W. F., Evans, J. O., & Reed, S. C. (1976, 01). Effects of Glyphosate on Chloroplast Ultrastructure of Quackgrass Mesophyll Cells. Weed Science, 24(1), 22-25. doi:10.1017/s0043174500065346
23. Cherry, T. (2014, 03). Toward a New Climate Agreement. doi:10.4324/9780203080009
24. Chuai, X. W., Huang, X. J., Wang, W. J., & Bao, G. (2012, 09). NDVI, temperature and precipitation changes and their relationships with different vegetation types during 1998-2007 in Inner Mongolia, China. International Journal of Climatology, 33(7), 1696-1706. doi:10.1002/joc.3543
25. Civeira, G. (2019, 02). Introductory Chapter: Soil Moisture. Soil Moisture. doi:10.5772/intechopen.83603
26. Clarke, A. J., & Gorder, S. V. (2001, 02). ENSO prediction using an ENSO trigger and a proxy for Western Equatorial Pacific Warm Pool Movement. Geophysical Research Letters, 28(4), 579-582. doi:10.1029/2000gl012201
27. Comaniciu, D. (2002). Bayesian Kernel Tracking. Lecture Notes in Computer Science Pattern Recognition, 438-445. doi:10.1007/3-540-45783-6_53
28. Cooper, M., Woodruff, D., Phillips, I., Basford, K., & Gilmour, A. (2001, 01). Genotype-by-management interactions for grain yield and grain protein concentration of wheat. Field Crops Research, 69(1), 47-67. doi:10.1016/s0378-4290(00)00131-3

29. Day, C., & Hanuschak, G. (2004, 10). Crop Area Estimation, Landsat Data Analysis. Encyclopedia of Statistical Sciences. doi:10.1002/0471667196.ess5079
30. Decker, M. (2008, 10). Agriculture and Agricultural Technology. Oxford Handbooks Online. doi:10.1093/oxfordhb/9780199252466.013.0036
31. Denègre, J. (1994, 01). Technological Progress In Geographical Research: Recent Developments In Satellite Remote Sensing And Geographic Information Systems. Mapping Sciences and Remote Sensing, 31(1), 3-12. doi:10.1080/07493878.1994.10641948
32. Dong, Y., Wang, J., Li, C., Wang, Q., & Huang, W. (2012, 06). Estimating the Leaf Area Index of Winter Wheat Canopies with Crop Growth Model CERES-Wheat. 2012 2nd International Conference on Remote Sensing, Environment and Transportation Engineering. doi:10.1109/rsete.2012.6260560
33. Drake, L. (1989, 12). Swedish agriculture at a turning point. Agriculture and Human Values, 6(1-2), 117-126. doi:10.1007/bf02219428
34. Drosdowsky, W. (2006). Statistical prediction of ENSO (Nino 3) using sub-surface temperature data. Geophysical Research Letters, 33(3). doi:10.1029/2005gl024866
35. Ehlers, W., & Goss, M. (n.d.). Water stress in plants. Water Dynamics in Plant Production, 209-223. doi:10.1079/9781780643816.0209
36. Elrick, D. E., & Bowman, D. H. (1964, 05). Note on an Improved Apparatus for Soil Moisture Flow Measurements. Soil Science Society of America Journal, 28(3), 450-453. doi:10.2136/sssaj1964.03615995002800030045x
37. Erickson, D. J., & Eaton, B. E. (1993, 04). Global biogeochemical cycling estimates with CZCS satellite data and general circulation models. Geophysical Research Letters, 20(8), 683-686. doi:10.1029/93gl00210
38. Fitzpatrick, G. E., & Verkade, S. D. (1989, 09). Using a Programmable Calculator for Determining Potential Evapotranspiration Rates. Journal of Environmental Horticulture, 7(3), 118-121. doi:10.24266/0738-2898-7.3.118
39. Franco, A. A., & Munns, D. N. (1982). Plant assimilation and nitrogen cycling. Nitrogen Cycling in Ecosystems of Latin America and the Caribbean, 1-13. doi:10.1007/978-94-009-7639-9_1

40. Franklin, A. G., & Coronado, P. L. (2010, 07). Prospects of new real-time radar applications for environmental remote sensing. 2010 IEEE International Geoscience and Remote Sensing Symposium. doi:10.1109/igarss.2010.5648803

41. Fraser, L. H. (2019, 03). Production changes in response to climate change. Grasslands and Climate Change, 82-97. doi:10.1017/9781108163941.007

42. Fu, C., Fletcher, J., & Slutz, R. (1983, 07). The Structure of the Asian Monsoon Surface Wind Field over the Ocean. Journal of Climate and Applied Meteorology, 22(7), 1242-1252. doi:10.1175/1520-0450(1983)0222.0.co;2

43. Garzelli, A., Capobianco, L., & Nencini, F. (2009). On the effects of pan-sharpening to target detection. 2009 IEEE International Geoscience and Remote Sensing Symposium. doi:10.1109/igarss.2009.5418022

44. Gates, D. M. (1990, 07). Climate change and the response of forests. International Journal of Remote Sensing, 11(7), 1095-1107. doi:10.1080/01431169008955082

45. Giampietro, M. (2013, 03). Energy Analysis for a Sustainable Future. doi:10.4324/9780203107997

46. Godfray, H. C. (2014, 05). How can 9–10 Billion People be Fed Sustainably and Equitably by 2050? Is the Planet Full?, 104-120. doi:10.1093/acprof:oso/9780199677771.003.0006

47. Granger, S., & Lizumi, H. (2001). Water quality measurement methods for seagrass habitat. Global Seagrass Research Methods, 393-406. doi:10.1016/b978-044450891-1/50021-9

48. Halide, H. (2017). Seasonal ENSO forecasting: Where does a simple model stand amongst other operational ENSO models? doi:10.1063/1.4973111

49. Hanari, K. Z. (2000, 04). Transformation of Contour Maps to Digital terrain Model (DTM). Journal of Zankoy Sulaimani - Part A, 3(1), 93-111. doi:10.17656/jzs.10056

50. Hatfield, J., & Walthall, C. (2014). Climate Change: Cropping System Changes and Adaptations. Encyclopedia of Agriculture and Food Systems, 256-265. doi:10.1016/b978-0-444-52512-3.00003-6

51. Headey, D. (2011, 04). Rethinking the global food crisis: The role of trade shocks. Food Policy, 36(2), 136-146. doi:10.1016/j.foodpol.2010.10.003

52. Hearn, A., & Bange, M. (2002, 10). SIRATAC and CottonLOGIC: Persevering with DSSs in the Australian cotton industry. Agricultural Systems, 74(1), 27-56. doi:10.1016/s0308-521x(02)00019-7
53. Hearn, A. (1994, 01). OZCOT: A simulation model for cotton crop management. Agricultural Systems, 44(3), 257-299. doi:10.1016/0308-521x(94)90223-3
54. Hill, B., & Ray, D. (1987). Incomes and Agriculture. Economics for Agriculture, 295-324. doi:10.1007/978-1-349-18782-9_12
55. Hillel, D. (2003). Irrigation and Water-Use Efficiency. Introduction to Environmental Soil Physics, 407-425. doi:10.1016/b978-012348655-4/50022-8
56. Hoogenboom, G. (2000, 06). Contribution of agrometeorology to the simulation of crop production and its applications. Agricultural and Forest Meteorology, 103(1-2), 137-157. doi:10.1016/s0168-1923(00)00108-8
57. Hook, S. C., Bone, G. T., & Fearn, T. (1982, 07). The conditioning of wheat. The effect of increasing wheat moisture content on the milling performance of uk wheats with reference to wheat texture. Journal of the Science of Food and Agriculture, 33(7), 655-662. doi:10.1002/jsfa.2740330711
58. Houérou, H. N. (2009). Bioclimatology and Biogeography of Africa. doi:10.1007/978-3-540-85192-9
59. Hume, D. (1995, 07). Pasture management: Technology for the 21st century. Agriculture, Ecosystems & Environment, 54(3), 224-225. doi:10.1016/0167-8809(95)90028-4
60. Hwang, E., Cho, S., & Kwon, H. (2004, 09). A Possible Role of Trehalose as a Regulatory Molecule in Plant Drought Resistance. Korean Journal of Environmental Agriculture, 23(3), 123-127. doi:10.5338/kjea.2004.23.3.123
61. Jones, T. (2020). The (Un)Sustainability of Manual Sugarcane Harvesting. Sugarcane Labor Migration in Brazil, 97-113. doi:10.1007/978-3-030-35671-2_5
62. Jørgensen, E. J. (1978, 11). Alternatives for Danish agriculture. Food Policy, 3(4), 308-312. doi:10.1016/0306-9192(78)90061-1
63. Kaakeh, W. (1997). The response of German Cockroaches to several toxic bait formulation. Emirates Journal of Food and Agriculture, 9(1), 1. doi:10.9755/ejfa.v9i1.5235

64. Kadkhoda, P., Blaschke, H., Kohlhaas, J., & Ristau, D. (2000, 11). Investigations of transmittance and reflectance in the DUV/VUV spectral range. Optical Metrology Roadmap for the Semiconductor, Optical, and Data Storage Industries. doi:10.1117/12.405831
65. Kang, J., Fernández-Beltrán, R., Ye, Z., Tong, X., Ghamisi, P., & Plaza, A. (2020, 08). High-Rankness Regularized Semi-Supervised Deep Metric Learning for Remote Sensing Imagery. Remote Sensing, 12(16), 2603. doi:10.3390/rs12162603
66. Kato, S. (2019). Assessing Negative and Positive Phototropism in Lianas. Methods in Molecular Biology Phototropism, 19-26. doi:10.1007/978-1-4939-9015-3_2
67. Kennedy, P., Leonforte, A., & Butsch, M. (2015, 07). Plant Breeding for Biological Nitrogen Fixation: A Review. Biological Nitrogen Fixation, 1071-1076. doi:10.1002/9781119053095.ch106
68. Keulen, H. (2013). Simulation Models crop/cropping management simulation models as Tools for Crop Management crop/ cropping management. Sustainable Food Production, 1459-1476. doi:10.1007/978-1-4614-5797-8_300
69. Khan, Y., Lawson, D., & Tucker, R. (1997, 10). Simple models of spectral radiative properties of carbon dioxide. International Journal of Heat and Mass Transfer, 40(15), 3581-3593. doi:10.1016/s0017-9310(96)00390-0
70. Kim, J., Kim, S., & Lee, Y. (2010, 03). SimTag: Exploiting tag bits similarity to improve the reliability of the data caches. 2010 Design, Automation & Test in Europe Conference & Exhibition (DATE 2010). doi:10.1109/date.2010.5456917
71. Kingwell, R., & Squibb, L. (2015). The role and value of combining dual-purpose crops and lucerne in a mixed-enterprise farming system. Crop and Pasture Science, 66(4), 399. doi:10.1071/cp14085
72. Knipling, E. B. (1970, 06). Physical and physiological basis for the reflectance of visible and near-infrared radiation from vegetation. Remote Sensing of Environment, 1(3), 155-159. doi:10.1016/s0034-4257(70)80021-9
73. Kolářová, M., & Hamouz, P. (2016, 04). Introduction to weed mapping methodology. Atlas of Weed Mapping, 425-427. doi:10.1002/9781118720691.ch41

74. Landsea, C. W. (2005, 12). Hurricanes and global warming. Nature, 438(7071). doi:10.1038/nature04477
75. Larcher, W. (1965). The Influence of Water Stress on the Relationship between Co2 Uptake and Transpiration. Water Stress in Plants, 184-194. doi:10.1007/978-94-010-3593-4_18
76. Lawes, R., Oliver, Y., & Robertson, M. (2009, 09). Integrating the effects of climate and plant available soil water holding capacity on wheat yield. Field Crops Research, 113(3), 297-305. doi:10.1016/j.fcr.2009.06.008
77. Leport, L., Turner, N. C., Davies, S., & Siddique, K. (2006, 04). Variation in pod production and abortion among chickpea cultivars under terminal drought. European Journal of Agronomy, 24(3), 236-246. doi:10.1016/j.eja.2005.08.005
78. Li, Z., Dong, Q., & Cao, G. (2005, 01). The photoelectrical measurement technique research for optical system transmittivity. Optoelectronic Devices and Integration. doi:10.1117/12.575057
79. Lizarazo, C., Stoddard, F., Mäkelä, P., & Santanen, A. (2010, 01). Genetic variability in the physiological responses of Andean lupin to drought stress. Suomen Maataloustieteellisen Seuran Tiedote, (26), 1-5. doi:10.33354/smst.76862
80. Lobell, D. (2009, 09). Crop Responses to Climate: Time-Series Models. Climate Change and Food Security Advances in Global Change Research, 85-98. doi:10.1007/978-90-481-2953-9_5
81. Loeb, V. J., Hofmann, E. E., Klinck, J. M., Holm-Hansen, O., & White, W. B. (2008, 11). ENSO and variability of the Antarctic Peninsula pelagic marine ecosystem. Antarctic Science, 21(2), 135-148. doi:10.1017/s0954102008001636
82. Longuenesse, J., & Leonardi, C. (1994, 08). Some Ecophysiological Indicators Of Salt Stress In Greenhouse Tomato Plants. Acta Horticulturae, (366), 461-468. doi:10.17660/actahortic.1994.366.57
83. Lovell, J., Haverd, V., Jupp, D., & Newnham, G. (2012, 06). The Canopy Semi-analytic Pgap And Radiative Transfer (CanSPART) model: Validation using ground based lidar. Agricultural and Forest Meteorology, 158-159, 1-12. doi:10.1016/j.agrformet.2012.01.020
84. Löw, F., Prishchepov, A., Waldner, F., Dubovyk, O., Akramkhanov, A., Biradar, C., & Lamers, J. (2018, 01). Mapping Cropland Abandonment in the Aral Sea Basin with MODIS Time Series. Remote Sensing, 10(2), 159. doi:10.3390/rs10020159

85. Mann, K., & Lazier, J. (2013, 06). The Oceans and Global Climate Change: Physical and Biological Aspects. Dynamics of Marine Ecosystems, 390-422. doi:10.1002/9781118687901.ch10

86. Masson, D., & Knutti, R. (2011, 06). Spatial-Scale Dependence of Climate Model Performance in the CMIP3 Ensemble. Journal of Climate, 24(11), 2680-2692. doi:10.1175/2011jcli3513.1

87. Mavi, H. S., & Tupper, G. J. (2004, 04). Agrometeorology. doi:10.1201/9781482277999

88. Mcmurtrie, R. E., Landsberg, J. J., & Linder, S. (1989). Research Priorities in Field Experiments on Fast-Growing Tree Plantations: Implications of a Mathematical Production Model. Biomass Production by Fast-Growing Trees, 181-207. doi:10.1007/978-94-009-2348-5_11

89. Meek, S., Goulding, J., & Priestnall, G. (2013, 04). The Influence of Digital Surface Model Choice on Visibility-based Mobile Geospatial Applications. Transactions in GIS, 17(4), 526-543. doi:10.1111/tgis.12021

90. Miller, M., Anderson, M., Francis, C., Kruger, C., Barford, C., Park, J., & Mccown, B. (2013, 09). Critical Research Needs for Successful Food Systems Adaptation to Climate Change. Journal of Agriculture, Food Systems, and Community Development, 1-15. doi:10.5304/jafscd.2013.034.016

91. Muchero, W., Ehlers, J. D., & Roberts, P. A. (2008, 03). Seedling Stage Drought-Induced Phenotypes and Drought-Responsive Genes in Diverse Cowpea Genotypes. Crop Science, 48(2), 541-552. doi:10.2135/cropsci2007.07.0397

92. Myers, D. R. (2013). Solar Radiation solar radiation for Solar Energy Utilization solar radiation for solar energy utilization. Solar Energy, 584-607. doi:10.1007/978-1-4614-5806-7_450

93. Nebot, E., Casanueva, J., Casanueva, T., Fernández-Bastón, M., & Sales, D. (2006, 11). In situ experimental study for the optimization of chlorine dosage in seawater cooling systems. Applied Thermal Engineering, 26(16), 1893-1900. doi:10.1016/j.applthermaleng.2006.02.001

94. Negri, R. G., Sant'anna, S. J., & Dutra, L. V. (2011, 07). Semi-supervised remote sensing image classification methods assessment. 2011 IEEE International Geoscience and Remote Sensing Symposium. doi:10.1109/igarss.2011.6049831

95. Nijsse, F., Cox, P., Williamson, M., & Huntingford, C. (2018, 12). Relationship between decadal climate variability and climate sensitivity. doi:10.1002/essoar.10500090.1
96. Nittel, S. (2009, 07). A Survey of Geosensor Networks: Advances in Dynamic Environmental Monitoring. Sensors, 9(7), 5664-5678. doi:10.3390/s90705664
97. OECD-FAO Agricultural Outlook 2000. (2000, 04). OECD-FAO Agricultural Outlook. doi:10.1787/agr_outlook-2000-en
98. O'neill, B. C., Oppenheimer, M., & Petsonk, A. (2006, 01). Interim targets and the climate treaty regime. Climate Policy, 5(6), 639-645. do i:10.1080/14693062.2006.9685584
99. Otter, S., & Ritchie, J. T. (1985). Validation of the Ceres-Wheat Model in Diverse Environments. Wheat Growth and Modeling, 307-310. doi:10.1007/978-1-4899-3665-3_28
100. Pal, M., & Maity, R. (2020, 03). Spatially-Varying Statistical Soil Moisture Profile Model by Coupling Memory and Forcing using Hydrologic Soil Groups to Estimate Vertical Soil Moisture Profile. doi:10.5194/egusphere-egu2020-6856
101. Patwa, N., & Penning, B. W. (2020). Environmental Impact on Cereal Crop Grain Damage from Pre-harvest Sprouting and Late Maturity Alpha-Amylase. Sustainable Agriculture in the Era of Climate Change, 23-41. doi:10.1007/978-3-030-45669-6_2
102. Pedgley, D. (2009, 12). Temperature and heat. Weather, 64(12), 340-340. doi:10.1002/wea.517
103. Peter, P. (2011, 04). Image Segmentation by Autoregressive Time Series Model. Image Segmentation. doi:10.5772/15570
104. Peters, G., Fischer, B., Münster, H., Clemens, M., & Wagner, A. (2005, 12). Profiles of Raindrop Size Distributions as Retrieved by Microrain Radars. Journal of Applied Meteorology, 44(12), 1930-1949. doi:10.1175/jam2316.1
105. Peterson, M. J., Hall, D. M., Feldpausch-Parker, A. M., & Peterson, T. R. (2010, 02). Obscuring Ecosystem Function with Application of the Ecosystem Services Concept. Conservation Biology, 24(1), 113-119. doi:10.1111/j.1523-1739.2009.01305.x
106. Piccioni, N. B. (2015, 11). Agriculture in Nicaragua. doi:10.1596/25978
107. Pitts, D., Badhwar, G., & Feiveson, A. (n.d.). Comparison Of Measured C-band Scattering Coefficients With Model Predictions As A Function

Of Leaf Area Index And Biomass. International Geoscience and Remote Sensing Symposium, 'Remote Sensing: Moving Toward the 21st Century'. doi:10.1109/igarss.1988.569443

108. Pollard, J. M., & Gatti, R. A. (2009, 08). Clinical Radiation Sensitivity With DNA Repair Disorders: An Overview. International Journal of Radiation Oncology*Biology*Physics, 74(5), 1323-1331. doi:10.1016/j.ijrobp.2009.02.057
109. Price, J. C. (1991, 02). Beltsville symposium XV remote sensing for agriculture. Remote Sensing of Environment, 35(2-3), 77. doi:10.1016/0034-4257(91)90001-m
110. Probert, M., Dimes, J., Keating, B., Dalal, R., & Strong, W. (1998, 01). APSIM's water and nitrogen modules and simulation of the dynamics of water and nitrogen in fallow systems. Agricultural Systems, 56(1), 1-28. doi:10.1016/s0308-521x(97)00028-0
111. Qiao, C., Daneshfar, B., Davidson, A., Jarvis, I., Liu, T., & Fisette, T. (2014, 07). Integration of optical and polarimetric SAR imagery for locally accurate crop classification. 2014 IEEE Geoscience and Remote Sensing Symposium. doi:10.1109/igarss.2014.6946718
112. Quan, N. H., & Meon, G. (2014, 05). Nutrient Dynamics During Flood Events in Tropical Catchments: A Case Study in Southern Vietnam. CLEAN - Soil, Air, Water, 43(5), 652-661. doi:10.1002/clen.201300264
113. Quiring, S. M. (2009, 01). Monitoring Drought: An Evaluation of Meteorological Drought Indices. Geography Compass, 3(1), 64-88. doi:10.1111/j.1749-8198.2008.00207.x
114. R., D. (2012, 03). Using Geographically Robust Models of Insect Phenology in Forestry. Phenology and Climate Change. doi:10.5772/35710
115. Raman, R., Diffey, S., Barbulescu, D. M., Coombes, N., Luckett, D., Salisbury, P., . . . Raman, H. (2020, 03). Genetic and physical mapping of loci for resistance to blackleg disease in canola (Brassica napus L.). Scientific Reports, 10(1). doi:10.1038/s41598-020-61211-y
116. Rao, A., & Boken, V. K. (2005, 05). Monitoring and Managing Agricultural Drought in India. Monitoring and Predicting Agricultural Drought. doi:10.1093/oso/9780195162349.003.0034
117. Rayne, S., Rayne, S., Forest, K., & Friesen, K. (2011, 02). Projected Climate Change Impacts on Grape Growing in the Okanagan Valley, British Columbia, Canada. Nature Precedings. doi:10.1038/npre.2011.3162.2

118. Riedell, W. E., & Blackmer, T. M. (1999, 11). Leaf Reflectance Spectra of Cereal Aphid-Damaged Wheat. Crop Science, 39(6), 1835-1840. doi:10.2135/cropsci1999.3961835x
119. Rikkinen, J. (2007). Relations Between Cyanobacterial Symbionts in Lichens and Plants. Microbiology Monographs Prokaryotic Symbionts in Plants, 265-270. doi:10.1007/7171_2007_102
120. Robertshaw, P., & Taylor, D. (2000, 03). Climate Change And The Rise Of Political Complexity In Western Uganda. The Journal of African History, 41(1), 1-28. doi:10.1017/s0021853799007653
121. Rosenzweig, C. (2001). Climate Impacts. International Encyclopedia of the Social & Behavioral Sciences, 2003-2010. doi:10.1016/b0-08-043076-7/04130-9
122. Ruschel, A. P., & Vose, P. B. (1982). Nitrogen cycling in sugarcane. Nitrogen Cycling in Ecosystems of Latin America and the Caribbean, 139-146. doi:10.1007/978-94-009-7639-9_12
123. Sabins, F. F. (2004). Remote Sensing for Energy Resources. Encyclopedia of Energy, 291-300. doi:10.1016/b0-12-176480-x/00272-2
124. Saha, T., Nithya, C., & Ray, S. (2015, 01). Climate Change: Impact on Pollinators' Biodiversity in Vegetable Crops. Climate Dynamics in Horticultural Science, Volume Two, 327-340. doi:10.1201/b18038-21
125. Sakar, D., Durutan, N., & Meyveci, K. (1988). Factors which limit the productivity of cool season food legumes in Turkey. World Crops: Cool Season Food Legumes Current Plant Science and Biotechnology in Agriculture, 137-145. doi:10.1007/978-94-009-2764-3_14
126. Sapaty, P. S. (2017). Spatial Grasp Model. Managing Distributed Dynamic Systems with Spatial Grasp Technology Studies in Computational Intelligence, 31-56. doi:10.1007/978-3-319-50461-2_3
127. Schmidt, G., Jenkerson, C. B., Masek, J., Vermote, E., & Gao, F. (2013). Landsat ecosystem disturbance adaptive processing system (LEDAPS) algorithm description. Open-File Report. doi:10.3133/ofr20131057
128. Sinharoy, S., Kryvoruchko, I. S., Pislariu, C. I., González-Guerrero, M., Benedito, V. A., & Udvardi, M. K. (2015, 07). Functional Genomics of Symbiotic Nitrogen Fixation in Legumes with a Focus on Transcription Factors and Membrane Transporters. Biological Nitrogen Fixation, 823-836. doi:10.1002/9781119053095.ch82

129. Skauli, T. (2009, 05). Sensor-informed representation of hyperspectral images. Algorithms and Technologies for Multispectral, Hyperspectral, and Ultraspectral Imagery XV. doi:10.1117/12.819491

130. Smith, D. S. (1964, 09). Ovarioles and Developing Eggs in Grasshoppers. The Canadian Entomologist, 96(9), 1255-1258. doi:10.4039/ent961255-9

131. Sobrino, J., Li, Z., Stoll, M., & Becker, F. (1994, 03). Improvements in the split-window technique for land surface temperature determination. IEEE Transactions on Geoscience and Remote Sensing, 32(2), 243-253. doi:10.1109/36.295038

132. Sonobe, R., & Wang, Q. (2017, 02). Towards a Universal Hyperspectral Index to Assess Chlorophyll Content in Deciduous Forests. Remote Sensing, 9(3), 191. doi:10.3390/rs9030191

133. Späth, H. (1992). Linear Lp Regression with Linear Constraints. Mathematical Algorithms for Linear Regression, 217-248. doi:10.1016/b978-0-12-656460-0.50011-2

134. Sørensen, R., Zinko, U., & Seibert, J. (2006, 02). On the calculation of the topographic wetness index: Evaluation of different methods based on field observations. Hydrology and Earth System Sciences, 10(1), 101-112. doi:10.5194/hess-10-101-2006

135. Sridhar, V. (2016). CLIMEX based potential distribution ofTuta absoluta(Meyrick) (Lepidoptera: Gelechiidae) under climate change. 2016 International Congress of Entomology. doi:10.1603/ice.2016.109123

136. Steduto, P., Çetinkökü, Ö, Albrizio, R., & Kanber, R. (2002, 05). Automated closed-system canopy-chamber for continuous field-crop monitoring of CO2 and H2O fluxes. Agricultural and Forest Meteorology, 111(3), 171-186. doi:10.1016/s0168-1923(02)00023-0

137. Stehlík, M., Stehlíková, S., & Torres, S. (2016). Understanding water extremes with caution. doi:10.1063/1.4951814

138. Stern, R. (2010). Statistics and Agrometeorology: Introductory Issues and Cases. Applied Agrometeorology, 939-947. doi:10.1007/978-3-540-74698-0_110

139. Sterner, T., & Hammer, H. (2005, 07). Designing instruments for climate policy. Emissions Trading for Climate Policy, 17-36. doi:10.1017/cbo9780511493485.003

140. Stigter, C., Sivakumar, M., & Rijks, D. (2000, 06). Agrometeorology in the 21st century: Workshop summary and recommendations on needs and perspectives. Agricultural and Forest Meteorology, 103(1-2), 209-227. doi:10.1016/s0168-1923(00)00113-1
141. Stone, M. C., Hotchkiss, R. H., & Stone, A. B. (2005, 07). Climate Change Impacts on Missouri River Basin Water Yields: The Influence of Temporal Scales. Impacts of Global Climate Change. doi:10.1061/40792(173)461
142. Stow, D., Petersen, A., Rogan, J., & Franklin, J. (2007, 03). Mapping Burn Severity of Mediterranean-Type Vegetation Using Satellite Multispectral Data. GIScience & Remote Sensing, 44(1), 1-23. doi:10.2747/1548-1603.44.1.1
143. Stępkowski, T., Banasiewicz, J., Granada, C., Andrews, M., & Passaglia, L. (2018, 03). Phylogeny and Phylogeography of Rhizobial Symbionts Nodulating Legumes of the Tribe Genisteae. Genes, 9(3), 163. doi:10.3390/genes9030163
144. Streeter, J. G., Peters, N. K., Salminen, S. O., Pladys, D., & Zhaohua, P. (1995, 03). Fate of Nodule-Specific Polysaccharide Produced by Bradyrhizobium japonicum Bacteroids. Plant Physiology, 107(3), 857-864. doi:10.1104/pp.107.3.857
145. Sung, J., & Seo, S. (2018, 10). Estimation of River Management Flow Considering Stream Water Deficit Characteristics. Water, 10(11), 1521. doi:10.3390/w10111521
146. Szpakowski, D., & Jensen, J. (2019, 11). A Review of the Applications of Remote Sensing in Fire Ecology. Remote Sensing, 11(22), 2638. doi:10.3390/rs11222638
147. Thenkabail, P. S. (2008, 11). Water productivity mapping methods using remote sensing. Journal of Applied Remote Sensing, 2(1), 023544. doi:10.1117/1.3033753
148. Thornton, P. (1993, 01). Risk in agriculture. Agricultural Systems, 43(1), 112-113. doi:10.1016/0308-521x(93)90096-k
149. Tilman, D. (2019, 05). Biodiversity and agriculture. Agricultural Resilience, 39-59. doi:10.1017/9781107705555.004
150. Tong, N., Li, L., & Lu, X. (2015). Maximum-Likelihood Classification for MPSK with Compressive Samplings. Proceedings of the First International Conference on Information Sciences, Machinery, Materials and Energy. doi:10.2991/icismme-15.2015.411

151. Tourre, Y. M., & White, W. B. (2005, 04). Evolution of the ENSO signal over the tropical Pacific-Atlantic domain. Geophysical Research Letters, 32(7). doi:10.1029/2004gl022128

152. Tsonis, A. A., Hunt, A. G., & Elsner, J. B. (2003, 10). On the relation between ENSO and global climate change. Meteorology and Atmospheric Physics, 84(3-4), 229-242. doi:10.1007/s00703-003-0001-7

153. Uehara, G., & Tsuji, G. Y. (1993). The IBSNAT project. Systems Approaches for Agricultural Development, 505-513. doi:10.1007/978-94-011-2842-1_31

154. Vermote, E., & Tanré, D. (1992, 04). Analytical expressions for radiative properties of planar Rayleigh scattering media, including polarization contributions. Journal of Quantitative Spectroscopy and Radiative Transfer, 47(4), 305-314. doi:10.1016/0022-4073(92)90149-x

155. Wallace, L., Lucieer, A., Malenovský, Z., Turner, D., & Vopěnka, P. (2016, 03). Assessment of Forest Structure Using Two UAV Techniques: A Comparison of Airborne Laser Scanning and Structure from Motion (SfM) Point Clouds. Forests, 7(12), 62. doi:10.3390/f7030062

156. Wang, S. (2016). Leaf area index. doi:10.4095/298882

157. White, R. E. (1997). Soil science: Raising the profile. Soil Research, 35(5), 961. doi:10.1071/s96090

158. Wilhite, D. A., & Pulwarty, R. S. (2017, 09). Drought and Water Crises. doi:10.1201/b22009

159. Willmer, C., & Fricker, M. (1996). The distribution of stomata. Stomata, 12-35. doi:10.1007/978-94-011-0579-8_2

160. Wollenberg, E. (2013, 03). Climate Change Mitigation and Agriculture. doi:10.4324/9780203144510

161. Yaghoobi, M., & Davies, M. E. (2010). Dictionary Learning for Sparse Representations: A Pareto Curve Root Finding Approach. Latent Variable Analysis and Signal Separation Lecture Notes in Computer Science, 410-417. doi:10.1007/978-3-642-15995-4_51

162. Zebiak, S. E. (2019, 08). International Conference on Climate Services - 5 – An introduction. Climate Services, 15, 100121. doi:10.1016/j.cliser.2019.100121

163. Zeng, X., Yang, P., & Niu, W. (2009, 05). Design of a Filter Based on Polynomial Interpolation and Local Fit. 2009 International Forum on Information Technology and Applications. doi:10.1109/ifita.2009.361

164. Zohary, M. (1963). On the Geobotanical Structure of Iran/ M. Zohary. doi:10.24157/arc_13414
165. Zohry, A., & Ouda, S. (2018). Crop Rotation Increases Land Productivity. Crop Rotation, 39-54. doi:10.1007/978-3-030-05351-2_3
166. Zuzulová, V., & Šiška, B. (2017, 05). Identification of Drought in Western Slovakia by Palmer Drought Severity Index (PDSI). Acta Regionalia Et Environmentalica, 14(1), 7-14. doi:10.1515/aree-2017-0002
167. Attri, S. D. (2020). Climate Services for Managing Climate Change Impacts on Agriculture. Global Climate Change: Resilient and Smart Agriculture, 301-314. doi:10.1007/978-981-32-9856-9_14
168. Baytorun, A. N., & Zaimoglu, Z. (2018, 03). Climate Control in Mediterranean Greenhouses. Climate Resilient Agriculture - Strategies and Perspectives. doi:10.5772/intechopen.72367
169. Mondal, M. (2020, 08). Smart Practices and Adaptive Technologies for Climate Resilient Agriculture. Advanced Agriculture. doi:10.30954/ndp-advagr.2020.1
170. Randall, N. P., & Smith, B. (2019, 11). Biodiversity. The Biology of Agroecosystems, 27-44. doi:10.1093/oso/9780198737520.003.0003

INDEX

Q

R

S

T

V

W

Y

Z